Massachusetts Institute of Technology

Vannevar Bush

science ■

the endless frontier

**A report to the President on a
Program for Postwar Scientific Research**

By Vannevar Bush

*Director of the
Office of Scientific Research and Development*

July 1945

Reprinted July 1960

National Science Foundation   Washington, D. C.

SF—60—40

Science — The Endless Frontier

"New frontiers of the mind are before us, and if they are pioneered with the same vision, boldness, and drive with which we have waged this war we can create a fuller and more fruitful employment and a fuller and more fruitful life."—

FRANKLIN D. ROOSEVELT.
November 17, 1944.

FOREWORD

The National Science Foundation has rendered a useful service in reprinting *Science, the Endless Frontier* as part of its tenth anniversary observance. The Report, as well as the studies that supported it, represents the collective efforts of a group of distinguished scientists and other scholars who brought their special experience and knowledge to bear on the problem of establishing a strong research and development effort in the postwar period. Their findings with respect to the relations of government to science and education merit a re-reading in the light of today's events. Dr. Waterman's Introduction constitutes an effective summary of the extent to which the recommendations of *Science, the Endless Frontier* have been realized during the fifteen years since it first appeared. I welcome the republication and hope that it will be a genuine service to all who have responsibilities for the national effort in scientific research and development.

VANNEVAR BUSH

TABLE OF CONTENTS

INTRODUCTION

ALAN T. WATERMAN

Director, National Science Foundation

On November 17, 1944, President Franklin D. Roosevelt addressed a letter to Vannevar Bush, director of the wartime Office of Scientific Research and Development, asking his advice as to how the lessons that had been learned from that experience could be applied in the days of peace that lay ahead. With the help and recommendations of four committees of distinguished scientists and other scholars, Dr. Bush set forth in clear and specific terms what he felt the relations of government to science should be and how these should be sustained. His report to the President was published in July 1945, under the imaginative title, *Science, the Endless Frontier*. The major recommendation was that a "National Research Foundation" should be established by the Congress to serve as a focal point for the support and encouragement of basic research and education in the sciences and for the development of national science policy. Five years later, in May, 1950, the Congress passed the National Science Foundation Act of 1950, bringing the new foundation into being.

On the occasion of the tenth anniversary of the Foundation's establishment, it seems appropriate to turn again to *Science, the Endless Frontier* and to attempt some assessment of the extent to which the objectives it set forth have been met. A re-reading impresses one with the perspicacity with which this remarkable document anticipated the major needs and problems relating to research and development in the postwar period. Although there have been shifts in emphasis since the report was written 15 years ago, its principles and its clear enunciation of the fundamental responsibility of the Federal Government in the area of research and development are as fresh and sound today as when they were written.

The original edition of *Science, the Endless Frontier* has long been out of print and the National Science Foundation is happy to make it available once more—not as an historical document, but as a classic expression of desirable relationships between government and science in the United States. Its usefulness and validity today are all the more remarkable when it is remembered that Dr. Bush and his advisers were of course quite unable to anticipate the specific developments that have most profoundly influenced our time, namely, the Korean war and the cold war, the missile and satellite race, the Soviet technological challenge, and the rapid acceleration of space research. Nor could Dr. Bush have estimated, in the final days of World War II, the full growth and direction of the atomic energy effort, including the large-scale programs and peaceful uses of nuclear energy. But he did anticipate in fullest measure that important developments would occur and that science and science education would be of immense importance in the postwar growth of the United States. The closing words of his Report were strongly prophetic: "On the wisdom with which we bring science to bear against the problems of the coming years depends in large measure our future as a nation."

Science and Government

Dr. Bush expressed the view that science is the proper concern of government but pointed out that the Government had only begun to utilize science in the Nation's welfare. He cited the absence within the Government of a body charged with formulating or executing national science policy and pointed out that there were no standing committees of the Congress devoted to this important subject. At the present time, science policy is constantly being made by the National Science Foundation with respect to basic research; by the President's Science Advisory Committee in matters in which the Chief Executive is responsible for direct action; and by the Federal Council on Science and Technology on coordination and planning that involve the interaction of the agencies of the Government concerned with research and development.

There are now three standing committees in the Congress whose concerns are directly related to science and technology: the Joint Committee on Atomic Energy; the Senate Committee on Aeronautical and Space Sciences; and the House Committee on Science and Astronautics. Twenty-four agencies within the Federal Government are responsible for the Government's obligation of funds for conduct of research and development, although nine agencies account for 99 per cent of the total.

The Importance of Basic Research

The principal focus of *Science, the Endless Frontier* is the importance of basic research. Of it, Dr. Bush said:

Basic research leads to new knowledge. It provides scientific capital. It creates the fund from which the practical applications of knowledge must be drawn. . . . Today, it is truer than ever that basic research is the pacemaker of technological progress. . . . A nation which depends upon others for its new basic scientific knowledge will be slow in its industrial progress and weak in its competitive position in world trade, regardless of its mechanical skill.

Dr. Bush viewed the publicly and privately supported colleges and universities and the endowed research institutes as the centers of basic research that must furnish both the new scientific knowledge and the trained research workers. He pointed out that if they were to meet the rapidly increasing demands of industry and government for new scientific knowledge, their basic research would have to be strengthened by the use of public funds.

Basic research is fundamental to all of the research and training needs which the Report considers. It is through basic research in biology, biochemistry and other sciences, for example, that the solutions to major disease problems are to be reached. Basic research is necessary to national defense if the United States is not to find itself fighting the next war with weapons merely improved from the last. Economic growth and the development of new products in industry are dependent upon rich resources of basic knowledge. And finally, knowledge of the methods and techniques of basic research is essential to the training and full development of skilled research investigators.

In the years since *Science, the Endless Frontier* was written, there has been an increased awareness on the part of the Government of the importance of basic research as shown by a steady trend upward in the amount of Federal funds for basic research and by an increase in the number and diversity of Government sources by which such support is furnished. However, the percentage of funds available for basic research has failed to increase in relation to total Federal funds for research and development—remaining somewhere between 6 and 7 per cent for a number of years.

In short, all the problems relating to the understanding and nurture of basic research in this country have not been solved.. The general public is still far from a true understanding of the nature of basic research and of the fundamental difference between science and technology. The evidence suggests that industry could profitably support a larger basic research effort both in its own laboratories and in the form of extramural support for colleges and universities. The relative proportion of Federal research and development funds between basic research and applied research and development has not achieved a completely desirable balance. These matters will be considered in somewhat greater detail further on under the discussion of the National Science Foundation.

Research Within the Government

Science, the Endless Frontier notes that research within the Government is an important part of our total research activity and urges that it be strengthened and expanded. In particular, it cites the need for revision of personnel practices and procedures in order to place the Government in a more advantageous position in competing with industries and the universities for first-class scientific talent.

Some progress has been made toward carrying out these recommendations. A series of legislative acts has created an excepted category for scientific personnel, authorized the Federal agencies to pay travel expenses to posts of duty, and provided opportunity for scientific and professional employees to take leave with pay for educational and training purposes. Changes in administrative attitudes have brought about an improved climate of opinion in Government laboratories which has resulted in increased opportunity and funds for Government scientists to engage in basic research. Publication in scientific journals is encouraged, and the payment of travel expenses to enable Government employees to attend scientific meetings is now rather generally accepted as right and necessary.

With respect to the organization of the administration of scientific activities within the Government, the Report declares:

In the Government the arrangement whereby the numerous scientific agencies form parts of larger departments has both advantages and disadvantages. But the present pattern is firmly established and there is much to be said for it. There is, however, a very real need for some measure of coordination of the common scientific activities of these agencies, both as to policies and budgets, and at present no such means exist.

The Report recommends:

A permanent Science Advisory Board should be created to consult with these scientific bureaus and to advise the executive and legislative branches of Government as to the policies and budgets of Government agencies engaged in scientific research.

The Report recommends that the board should be composed "of disinterested scientists who have no connection with the affairs of any Government agency."

These observations are of particular interest in view of the current debate over the need for a Department of Science and Technology. The coordination of common scientific activities, both as to policies and budget, is the responsibility for the newly established Federal Council on Science and Technology; and the advice and counsel of disinterested scientists is available to the President through his Science Advisory Committee.

It should be noted, however, that full attention to these matters was stimulated primarily by the Russian sputnik. Immediately after its successful launching, the post of Special Assistant to the President for Science and Technology was created; and the President's Science Advisory Committee—which had been established under the Office of Defense Mobilization in 1950—was reconstituted and placed directly under the President.

Industrial Research

The Bush Report approaches the issue of industrial research by stating directly: "The simplest and most effective way in which the Government can strengthen industrial research is to support basic research and to develop scientific talent." It goes on to point out, however, that one of the most important factors affecting the amount of industrial research is tax law, and it recommends that the Internal Revenue Code be amended to remove uncertainties in regard to the deductibility of research and development expenditures as current charges against net income.

The tax laws have now been changed, partially at least, to meet this particular problem. Among various legislative provisions designed to encourage business participation in private research ventures are Section 174 of the Internal Revenue Code of 1954, which permits business expenditures for research to be deducted from taxable income, and Section 9 of the Small Business Act of 1958, which encourages small business concerns to engage in joint research and development efforts.

The contributions of industrial research to our development as a nation are too obvious to require review. Furthermore research and development are themselves developing into a major industry for which the late Sumner Slichter coined the phrase, "industry of discovery." The importance of research to economic stabilization and growth is now almost universally recognized. In 1958 the National Science Foundation sponsored a conference on research and development and its impact on the economy. The impact of the conference itself was such that industrial officials who attended confessed afterwards that the conference had convinced them that they should not reduce research and development expenditures in the face of the 1958 recession. It is hoped that industry will continue to accord full support to basic research, both in its own laboratories, and, to the extent possible, through extramural support of basic research in the universities.

Organized labor is also developing an increasing awareness of the relation of research to the health and growth of the economy. In 1959 the AFL-CIO sponsored a conference on "Labor and Science in a Changing World." The conference acknowledged the inevitability of the technological progress and explored ways in which organized labor could meet the challenges and demands of the new technology.

Medical Research

Medical research is a point of major emphasis in *Science, the Endless Frontier*. An entire chapter, "The War Against Disease," is devoted to it and it was studied in great detail by one of the four advisory committees. Uppermost in the minds of Dr. Bush and his consultants were the impressive accomplishments of the military medical research and development effort and the absence of a specific agency for their continued support following the close of the war. Here again the emphasis was on basic studies. The Report observes:

It is wholly probable that progress in the treatment of cardiovascular disease, renal disease, cancer, and similar refractory diseases will be made as the result of fundamental discoveries in subjects unrelated to those diseases and perhaps entirely unexpected by the investigator. Further progress requires that the entire front of medicine and the underlying sciences of chemistry, physics, anatomy, biochemistry, physiology, pharmacology, bacteriology, pathology, parasitology, etc, be broadly developed.

Both Dr. Bush and his Medical Advisory Committee recommended action on the part of the Federal Government to initiate a support program for basic medical research in the medical schools and in the universities through grants for research and through fellowships. Dr. Bush recommended that the proposed program be administered by a "Division of Medical Research" of the "National Research Foundation"; the committee recommended that a second organization be established, to be called the National Foundation for Medical Research. Actually, both recommendations have been met by subsequent events, which resulted in both a division within the National Science Foundation that supports basic medical science (Division of Biological and Medical Sciences) and in a completely independent organization, the National Institutes of Health, which has far surpassed in its support programs anything that the Committee envisioned in the recommended Medical Research Foundation.

The two sets of recommendations did not differ greatly as to the amount of support that should be established at the initiation of the program—Bush recommending an initial start of $5 million a year extending upwards to perhaps $20 million a year at the end of five years; his Medical Advisory Committee recommending a start of approximately $5 to $7 million annually, with larger sums to follow as the program developed. The Committee urged the need for unrestricted grants, with support of fellowships and projects being of relatively less importance in their thinking.

A striking feature of postwar developments in the Government's support program for medical and health-related sciences has been the rapid rate of increment of funds. This is the result of the deep and continuing interest of the Congress in the progress of medical research. The National Institutes of Health has increased its obligations for research grants alone from $85,000 in 1945 (a year when the Bush Report suggests $5 to $7 million) to more than $155 million for grants and contracts in 1959. In addition, of course, the Institutes operate their own intramural research program at the Clinical Center and funds for this were around $45 million for 1959.

That both organizational recommendations have been met—that is, for a division of the Foundation and for a separate institutional organization—appears to have been a fortunate turn of affairs. The National Institutes of Health stresses research aimed at the care and cure of diseases, including basic research related to its mission, as defined by Executive Order 10512. The National Science Foundation, on the other hand, supports basic research in this area primarily for the purpose of advancing our knowledge and understanding of biological and medical fields. With more than one source of funds available from the Federal Government, scientists enjoy the broader base of support that is consistent with American tradition.

Although the U. S. Public Health Service and the National Science Foundation are the principal sources of funds for medical research, mention should also be made of the intramural programs of the Veterans Administration, the military services, and the medical research programs of the Atomic Energy Commission.

Military Research

With the civilian Office of Scientific Research and Development just bringing to a close its brilliantly successful program of wartime research on weapons and devices of warfare, and problems of military medicine, Dr. Bush felt that a certain amount of long-range scientific research on military problems should continue to be carried on in peacetime by a civilian group. Such research would complement research on the improvement of existing weapons which, he felt, could best be done within the military establishment. He therefore recommended that the new "National Research Foundation" should include a division of national defense. For this he contemplated a modest level of expenditures of $10 million for the first year, rising to $20 million by the end of the fifth year.

Here again, as in the case of medical research, the situation evolved in a way quite different from that originally visualized by Bush, but which has probably met the substance of his principal recommendations. A division of national defense was stricken from proposed legislation establishing a new agency largely because the delay had resulted in different measures being taken. The military services, who were well pleased with the civilian research performed in the universities under OSRD sponsorship, continued such arrangements with the universities by writing appropriate new contracts to continue the work started under OSRD auspices or to launch entirely new investigations. In ensuing years, many contracts of this type were entered into by the military services with a growing number of universities. The central laboratories originally associated with OSRD contracts, such as the Applied Physics Laboratory, Johns Hopkins University, the Radiation Laboratory at M.I.T., and the Jet Propulsion Laboratory of the California Institute of Technology, developed into the research centers, which, though supported by military funds, are operated by civilian scientists under civilian management.

In addition to applied research for the solution of immediate problems, the three services gradually expanded their research programs to include grants for basic research—in general related to their missions but often of a very fundamental nature. During the five-year period between the publication of *Science, the Endless Frontier* and the enactment of the National Science Foundation legislation in 1950, the Navy Department, through its Office of Naval Research (established by Congress in 1946) gave generous support to basic research in a wide variety of fields. Later, by order of their respective secretaries, a similar pattern was adopted by the Department of the Army, through its Office of Ordnance Research, and the Air Force, through the Air Force Office of Scientific Research.

The Bush thesis that "some research on military problems should be continued, in time of peace as well as in war, by civilians independently of the military establishment" has not been adequately tested because of the uncertain character of the peace that has existed since the close of World War II.

In general, however, it can be said that a substantial number of the Nation's top scientists, both within the Department of Defense and in outside institutions, are applying their talents to military problems with imagination and vigor.

International Relations in Science

With their long tradition of effective international cooperation in science, it is not surprising that a group of scientists should urge upon the Government the importance of a vital program for the continuing international exchange of scientific information, through both the medium of scientific literature and active participation in international conferences, symposia, and other forms of international collaboration in science.

Thus the Committee on Science and the Public Welfare recommended that scientific attachés be appointed to serve in certain selected United States embassies. "Such a post," observed the Committee, "would appear to be most important in countries such as Russia, where a great deal, if not all, of the scientific activity is controlled or directed by the government and where other channels of scientific communication have been greatly restricted for several years." This recommendation was reaffirmed by a special Department of State International Science Steering Committee in its report, *Science and Foreign Policy*, released in May 1950 at about the same time the National Science Foundation legislation was being enacted.

The Office of Science Adviser to the Secretary of State was established in a preliminary way in the summer of 1950 and the post of Science Adviser was formally filled as such in February 1951. During the first year, science attachés were assigned to London, Stockholm and Bern. The following year similar posts were added in Bonn and Paris.

Following the resignation of the Science Adviser in July 1953, the program was gradually permitted to lapse. Meanwhile, the scientific community, which felt that the program had made a definite contribution to international understanding and cooperation in science, pressed for a reactivation of the program—principally through the medium of editorial comment as expressed in various scientific journals. Under urging from the National Academy of Sciences, the National Science Foundation, and the President's Science Advisory Committee, the Department of State decided to renew and strengthen the program in July 1957.

The new Science Adviser took office in January 1958. In January 1959 the attaché program was again active, with attachés assigned to London, Paris, Stockholm, Bonn, Rome, Tokyo, New Delhi, Rio de Janeiro, and Buenos Aires. At the present time no real obstacle appears to exist to the fulfillment of the original Bush proposal that a scientific attaché in Moscow would be useful. In addition, the way appears to have been opened for better exchange between the U. S. and the U.S.S.R. of both scientific information and scientists under the Bronk-Nesmeyanov Agreement of July 1959.

As far as international scientific conferences are concerned, the situation is probably more satisfactory than at any time since the Bush Report was published. U. S. attendance at such meetings has been strengthened and placed on a more orderly basis through the National Science Foundation's authority to pay travel expenses of American scientists attending scientific meetings abroad and through the continued backing of the scientific unions by the National Academy of Sciences. In general, scientists are chosen to represent the U. S. by their peers acting usually through the scientific societies.

With respect to foreign scientists traveling to the U. S. for scientific meetings, there have been some improvements in the situation. The visa problem of recent years has been greatly alleviated. The current problem, which is a very real one for science but which transcends scientific considerations, is the problem of recognition and non-recognition of certain nations. The international scientific community operates without regard to political considerations and establishes its meetings and selects its delegates solely on the basis of their scientific qualifications. When these run head-on into political considerations involving the entrance of foreign nationals, there are, of course, knotty problems to be solved. At the present time, these problems appear to admit of no easy solution.

On the positive side, by far the largest and most impressive example of effective international cooperation in science was the International Geophysical Year of 1957-58. The U. S. was one of 66 nations participating in the

18-month period of intensive geophysical research. The scientific program was under the direction of the U. S. National Committee for the International Geophysical Year, National Academy of Sciences. Here again the Bush Report anticipated what was to come by citing the International Polar Year as an example of significant international scientific activity. The Report recommended that "the National Research Foundation be charged with the responsibility of participating in such international cooperative scientific enterprises as it deems desirable." The National Science Foundation secured and administered Government funds for U. S. participation in the International Geophysical Year to the extent of $43,500,000.

In the contemporary scene, international activities in science have necessarily widened to include political considerations. Through such mediums as the International Conference on the Peaceful Uses of Atomic Energy, the nations of the world are working to divert the powerful new forces of nuclear energy into constructive uses. Similarly, the nations may find it necessary in the common good to agree and cooperate on scientific and practical aspects of outer space research. The Antarctic Treaty under which twelve nations have agreed to preserve the Antarctic as a great scientific laboratory is a major landmark in international relations.

Renewal of Scientific Talent

In a chapter entitled "Renewal of Our Scientific Talent," Dr. Bush takes as a major premise the statement of James B. Conant that ". . . in every section of the entire area where the word *science* may properly be applied, the limiting factor is a human one. We shall have rapid or slow advance in this direction or in that depending on the number of really first-class men who are engaged in the work in question. . . . So in the last analysis the future of science in this country will be determined by our basic educational policy."

Dr. Bush and his advisory committee on education were concerned (1) with broadening the base from which students with scientific aptitude and talents could be drawn, and (2) with filling the wartime deficit in young scientists and engineers. They were concerned with quality and with the full operation of the democratic process. They felt that all boys and girls should be able to feel that, if they have what it takes, there is no limit to the opportunity. A ceiling should not be imposed on a young person's educational opportunities either by limited family means or negative family attitudes.

Science, the Endless Frontier also emphasized the importance of teaching in these words: "Improvement in the teaching of science is imperative; for students of latent scientific ability are particularly vulnerable to high school teaching which fails to awaken interest or to provide adequate instruction."

The specific recommendations of the Bush Report in the area of science education were for the establishment of a national program of science scholarships and science fellowships and for the subsequent enrollment of the recipients of these awards in a National Science Reserve upon which the Government could draw in times of emergency.

In the establishment and operation of the Foundation's program of education in the sciences, there has been fundamental and perhaps unanimous agreement with the Bush thesis. The methods and techniques by which these objectives are to be accomplished do not coincide at every point with the rather general proposals set forth in the Bush Report; nevertheless, I think it can be said that all the programs that the Foundation has initiated and supported have contributed in significant measure to the principal recommendation of Dr. Bush, namely, that the Nation's pool of scientific talent should be strengthened and improved.

In the very first year of operation with its total budget only $3.5 million, the Foundation awarded 575 predoctoral and postdoctoral fellowships. Over the ten-year period the fellowship program has been gradually expanded to include fellowships in other categories, and more than 12,000 fellowships in all categories have been awarded.

The Foundation has not embarked upon a program of scholarship support for a number of reasons, the principal one being the conviction of the National Science Board that an undergraduate program of scholarship support should not be limited to a particular field of science or even to science and engineering generally. The Foundation does, however, support several programs of a different type which provide to gifted students, at both the undergraduate and secondary-school levels, research experience and educational opportunities far beyond those afforded by the normal curriculum.

Financial assistance for undergraduate students was anticipated by Dr. Bush and his Committee. Although Public Law 346 (G.I. Bill of Rights) had been passed in 1944 and is mentioned at some length in *Science, the Endless Frontier*, its ultimate impact was not apparent at that time. The final summing up is impressive. Of the more than 7.5 million veterans who took advantage of this training, more than two million pursued courses in schools of higher learning. Almost 10 per cent of the total (744,000) pursued courses in scientific fields. The engineering profession attracted 45,000 and medicine and related courses more than 180,000. The remaining 113,000 who elected to study in the natural sciences were variously distributed among geology, chemistry, geography, metallurgy, physics, medicine, dentistry, and others.

About two million veterans of the Korean conflict received similar educational opportunities under the Veterans Readjustment Assistance Act of 1952. Engineering, medical, dental, and scientific fields attracted about a quarter million of these.

Other sources of financial aid for undergraduate students include the National Merit Scholarship Corporation, a nonprofit institution established and supported by philanthropic foundations and business organizations, and the National Defense Education Act of 1958, which provides for loans to students in institutions of higher education.

Dr. Bush's urgent plea that the generation in uniform should not be lost seems to have been abundantly answered. The evidence suggests also that the military services are making constructive efforts to utilize both draftees and officers in positions in which they can make use of specialized skills and training. The services also have interesting programs for continuing the advanced education of highly qualified men through such mediums as the Navy Postgraduate School and through direct subsidy of advanced education for military men in colleges and universities.

A comparison of the support levels for scientific personnel and education recommended in *Science, the Endless Frontier* and those that actually obtain is difficult. The Bush recommendation of $7 million for the first year, rising to $29 million by the fifth year, was based on an annual program of 6,000 undergraduate scholarships and 300 graduate fellowships. The National Science Foundation's obligations for scientific personnel, education and manpower, which in the early years were devoted largely to graduate fellowship support, totaled approximately $1.5 million the first year and $4 million about the fifth year. During this period, of course, Federal funds for education were also available through the G.I. Bill, through the fellowships of the National Institutes of Health, and the Atomic Energy Commission, as well as from other sources.

By 1960 the Foundation's obligations for scientific personnel and education totaled more than $65 million, of which more than half went for institutes to improve the teaching of mathematics and science principally in—but not limited to—the high school. The institutes program initiated by the Foundation on an experimental basis in 1953 appealed particularly to Congress and for several succeeding years funds have been specifically appropriated by Congress for this purpose.

A significant assessment of the impact and value of these programs is difficult at close range. A number of years, possibly a generation, will be required before we may be able to judge fairly the extent to which Federal-support programs have met their objectives.

In the National Science Foundation, quality rather than numbers has been stressed. We have felt that it was important for the whole broad rank and file of students to be made aware of the opportunities and intellectual satisfactions of science as well as other fields; it has seemed to us especially important that those with special aptitudes and ability from whatever walk of life should have the fullest opportunity for the realization of their talents.

The Foundation is trying to the extent possible to meet the problem at its source. It agrees fully with the Bush stress upon the importance of the teaching of science at the high school level. It has been apparent that in order to teach modern science effectively, teachers must not only be adequately trained themselves but must have the opportunity to work with up-to-date curriculums and course content and with proper laboratories and equipment.

Beginning with the work of the Physical Sciences Study Group at M.I.T., the Foundation is supporting studies looking toward the complete revision and up-dating of course content in physics, mathematics, chemistry, and biology. This work has included the preparation of new textbooks and teaching aids and the introduction of imaginative and stimulating new equipment.

It seems reasonable to assume that these constructive efforts must by their very nature influence for the better the teaching of science. Nevertheless, nothing that has been accomplished thus far provides reason for complacency. As a nation we still seem a long way from a universal understanding and appreciation for intellectual activity generally and probably will remain so until we attach roughly the same importance to academic achievement as we do, for example, to prowess in sports.

Reconversion

At the close of the war Dr. Bush and the scientific community generally were keenly aware of the volume and importance of the scientific information generated during the War and which had necessarily been subject to severe security restrictions. Of the medical information developed during the War, however, the greater part had remained unclassified and had been published. Dr. Bush expressed the view ". . . that most of the remainder of the classified scientific material should be released as soon as there is ground for belief that an enemy will not be able to turn it against us in this war."

On the whole, this problem seems to have been successfully met. Despite the enormous volume of material involved and the shortages of military and technical personnel qualified to rule on the security status of technical data, declassification has been steadily going on since the War. The Department of Defense through its Office of Declassification Policy and the Atomic Energy Commission through its Division of Classification are actively attacking this problem on a continuing basis. In the opinion of some who are dealing with this problem, more reports are being declassified than consumers can find time to read.

The latter point, of course, is related to the whole broad problem of the dissemination, storage, and retrieval of scientific information. This problem has received attention from the National Science Foundation since its inception, but because of limitation of funds, activities in this area were necessarily supported at a fairly low level until recently. Early in 1958 the President's Science Advisory Committee made a detailed study of what the Government should do to improve the flow of scientific information and thereby increase its utilization. As a result of the Committee's recommendations, the President directed that the scientific information activity of the National Science Foundation should be strengthened and expanded. At about the same time, under Title IX of Public Law 85-864 (the National Defense Education Act of 1958), the Foundation was authorized to establish a Science Information Service and also a Science Information Council, which would include in its membership outstanding scientists, information experts, and heads of Federal Bureaus and agencies that are directly concerned with the dissemination of information.

In a number of programs the Office of Science Information Service of the Foundation has sought to improve the dissemination of existing materials by helping to provide for prompt publication of research results, reference aids and information centers of various kinds, and translations of significant scientific papers in languages not widely understood by American scientists. In addition, the Foundation is supporting a slowly growing body of research directed to whole new approaches in various aspects of the information problem. Most of the research is concerned with exploration of ways of using machines to help with information processing tasks, such as the organization, storage and searching of scientific information and the translation of scientific publications from foreign languages into English. Before machines can process the texts of documents, however, for either mechanized information searching systems or mechanical translation systems, more precise knowledge of syntax and semantics is needed. Therefore, current research activities in these areas are extending our understanding of language with the expectation that ultimately machines will be able to handle linguistic data.

The recommendation of *Science, the Endless Frontier* that the National Research Foundation should include a Division of Publications and Scientific Collaboration has been substantively realized by the creation within the National Science Foundation of the Office of Science Information Service.

The National Science Foundation

In a final chapter labeled "The Means to the End," *Science, the Endless Frontier* recommends the establishment of a National Research Foundation, conceived as the principal means for carrying out the other major recommendations contained in the Report. The five years of legislative debate during which the scientific community urged upon Congress the importance of establishing a new foundation are history too familiar to require repetition here.

President Truman's veto of the bill that was finally passed by both Houses of Congress in 1947 was a major disappointment. The President's objections were directed toward the administrative structure of the new agency under which the director would be elected by a board, a provision which he felt would render it insufficiently responsive to the will of the people.

The Bill that was finally passed in May 1950 met the principal objections of the President by specifying that both the Director and the Members of the 24-member board should be appointed by the President. This unusual arrangement left over-all policy determination and program approval largely in the hands of the Board, with the Director reporting to the President, although serving *ex officio* on the Board and acting as its executive officer.

In 1958 the Board, through an *ad hoc* committee appointed for the purpose, reviewed the working relationship of the Director and the Board in the light of experience and noted that this relationship has been harmonious and constructive largely as a result of the excellent cooperation on the part of both. The Board noted further that each year of successful operation, built on a clear understanding on the part of each Board Member of his proper function, and upon wise statesmanship on the part of the Director and his associates gives assurance of continued success. The Board further observed that as each year passes a body of precedents for sound administrative procedures is being built up that may ultimately become an unwritten constitution which will prevail.

Some of the organizational anomaly of the Foundation was resolved in 1959 when Congress amended the National Science Foundation Act to permit the Board to delegate authority to the Director and its Executive Committee to approve grants and contracts in certain situations. The delegation of authority has since been implemented by Board action.

In other details, the structure of the National Science Foundation, as finally constituted, does not differ substantially from that proposed by *Science, the Endless Frontier*, except in the omission of a Division of National Defense. The Report proposed the following Divisions: Medical Research, Natural Science, National Defense, Scientific Personnel and Education, Division of Publications and Scientific Collaboration and appropriate staff offices.

The principal divisions of the Foundation are: Biology and Medicine; * Mathematical, Physical and Engineering Sciences; and Division of Scientific Personnel and Education. There is also an Office of Science Information Service, an Office of Special International Programs, an Office of Special Studies, as well as an administrative division.

In 1958 the National Science Board approved the establishment of an Office of Social Sciences, thus bestowing formal status on Foundation support of the social sciences which has been carried on on a limited basis since the early days of the Foundation. The social sciences support program which was undertaken on the strength of the permissive phrase in the Act "and other sciences" includes projects in the following fields: physical anthropology, functional archaeology, cultural anthropology, psycholinguistics, human ecology, demography, sociology, social psychology, economic and social geography, economics, history of science, and philosophy of science.

Thus in the National Science Foundation we have a functioning organization closely resembling in organization and design the National Research Foundation proposed by Dr. Bush.

Certainly in ideals and objectives it is a prototype of the institution envisioned in *Science, the Endless Frontier*. The really important question is, of course, to what extent has the Foundation been successful in serving the high purposes and needs for which it was created?

Dr. Bush enunciated five basic principles that should characterize an effective program of Government support for scientific research and education:

(1) Whatever the extent of support may be, there must be stability of funds over a period of years so that long-range programs may be undertaken.

(2) The agency to administer such funds should be composed of citizens selected only on the basis of their interest in and capacity to promote the work of the agency. They should be persons of broad interest in and understanding of the peculiarities of scientific research and education.

(3) The agency should promote research through contracts or grants to organizations outside the Federal Government. It should not operate any laboratories of its own.

(4) Support of basic research in the public and private colleges, universities, and research institutes must leave the internal control of policy, personnel, and the method and scope of the research to the institutions themselves. This is of the utmost importance.

(5) While assuring complete independence and freedom for the nature, scope, and methodology of research carried on in the institutions receiving public funds, and while retaining discretion in the allocation of funds among such institutions, the Foundation proposed herein must be responsible to the President and Congress.

(1) The Foundation is aware that continuity and stability are most important in the support of basic research. This can be achieved in two principal ways: one, by making the grant or contract for a term of years rather than for a single year and renewable; the other, by setting aside a revolving fund for the renewal of grants or contracts when the term expires, without reference to the annual appropriation. In the early years, budget limitations made it difficult for the Foundation to provide long-range support. As funds have become available, however, the amount and duration of the average grant has steadily increased. The value of the average research grant in fiscal year 1953 was $10,300, for an average duration of 1.9 years; and in fiscal year 1960 the average estimated value has risen to $30,500, with an average

---

* Under the terms of the Act biology and medical research were initially to be separate divisions. After consideration during the first year it was decided to combine these into a single division.

duration of 2.3 years. Individual grants are being made for as long as five years. Thus fuller support is being achieved, and with grants of longer duration the trend is toward increasing stability.

(2) The National Science Foundation has been extraordinarily fortunate in the calibre of the people who have manned its regular staff, as well as in the advisors and consultants who have served untiringly. The National Science Board, as prescribed by law, is composed of persons "eminent in the fields of the basic sciences, medical science, engineering, agriculture, education or public affairs; . . . selected solely on the basis of established records of distinguished service; and . . . so selected as to provide representation of the views of the scientific leaders in all areas of the Nation." The substantive divisions have statutory divisional committees of scientists eminent in their respective fields or specialties, and a similar committee for the Office of Social Sciences was recently named. At the program level there is an advisory panel for each program which advises and counsels the program director and provides assistance in the formulation of the program in that particular discipline. The regular Foundation staff is selected largely from college and university faculties—many serving on leave for the term of their appointment.

The Foundation maintains viable relationships with the scientific and educational communities which make possible constant interchange of views and information.

(3) The Foundation is permitted to operate no laboratories of its own. Early in its history the decision was made that the grant, in general, affords the most effective means of support for basic research. As a result of the Foundation's efforts, furthermore, the Eighty-Fifth Congress passed Public Law 934, which extends the grant-making authority to appropriate Federal agencies and permits them, also, in the case of basic research grants, to vest title to research equipment with the institution receiving the grant, provided such equipment is not needed for government purposes.

The need for major facilities for basic research purposes—not wholly anticipated by the Bush Report—has given rise to a situation in which the Foundation is supporting the construction and operation of such facilities by means of contracts with qualified organizations. For example, in astronomy, where urgent need exists for both photoelectric instruments and radiotelescopes, the Foundation is supporting two major facilities: the National Radio Astronomy Observatory at Green Bank, West Virginia, and the Kitt Peak National Observatory near Tucson, Arizona. The former is being operated by Associated Universities Incorporated, composed of nine eastern universities, which has had notable experience in the management of large-scale research in the operation of the Brookhaven National Laboratory. The Kitt Peak Observatory is being constructed and operated by the Association of Universities for Research in Astronomy, a group of eight universities with major astronomy departments, which was organized specifically for this purpose.

The Foundation has also contributed substantial support to other large-scale facilities for basic research, including high-speed computers, an oceanographic research vessel, and nuclear reactors.

An even more recent program, which is somewhat related to the facility program, is the development of graduate research laboratories under which the Foundation provides funds for the renovation and equipment of the research laboratories of graduate schools.

(4) In the operation of its program, the National Science Foundation has sought to hold to a minimum the burdens imposed upon academic institutions. Administrative requirements on grantees, fellows and contractors are the minimum consonant with accountability and responsibility for public funds. In the last analysis, however, the scientific and academic communities must be the final judge of the extent to which Federal support has been given without interference in internal affairs or burdensome controls. During its first ten years of operation the Foundation has had no serious complaints on this score.

(5) The Foundation has found its responsibilities to the President and the Congress in no wise incompatible with its independence and freedom of operation. Congress in its wisdom endowed the Foundation with an unusually broad charter. It is so broad, in fact, that the Foundation from time to time has had to place its own interpretation on its Act and to make policy decisions regarding what *not* to do. This wide latitude has enabled the Foundation to approach the immense and challenging problems of modern science in innovational and experimental ways.

The Director enjoys cordial working relationships with the Special Assistant to the President and with the President's Science Advisory Committee. Whenever circumstances require it, he has direct access to the President. The Director is a member of the National Aeronautics and Space Council, the Federal Council on Science and Technology, a consultant to the President's Science Advisory Committee and a member of the Defense Science Board.

The foregoing summary probably represents the extent to which we are able to comment on the success with which these five fundamentals have been met. A more complete judgment must await the perspective of history.

So far as the operations of the Foundation are concerned, these have been substantially covered in the course of commenting on the major recommenda tions of *Science, the Endless Frontier.* Upon examination, the Foundation s programs, particularly in the area of research support and education in the sciences, will be found to correspond closely with the principal recommenda tions of *Science, the Endless Frontier.*

An extremely troublesome and difficult problem is the Foundation's relation to the development of national science policy and to the evaluation and correlation functions. The National Science Foundation Act authorizes and directs the Foundation—

to develop and encourage the pursuit of a national policy for the promotion of basic research and education in the sciences;

. . . .

to evaluate scientific research programs undertaken by agencies of the Federal Government, and to correlate the Foundation's scientific research programs with those undertaken by individuals and by public and private research groups; . . .

The number and variety of Federal research programs prompted the Foundation at the outset to consider what should be the responsibilities of the several Federal agencies with respect to the support of extramural research and development in the sciences. After conferences by NSF staff members with the Bureau of the Budget and other agencies, the Foundation's primary recommendations were set forth in Executive Order 10521 of March 17, 1954. The Order states that the Foundation "shall . . . recommend to the President policies for the promotion and support of basic research and education in the sciences, including policies with respect to furnishing guidance toward defining the responsibilities of the Federal Government in the conduct and support of basic scientific research."

The Order further directs that the Foundation shall be increasingly responsible for the support of general-purpose basic research but recognizes, also, the importance and desirability of other agencies' conducting and supporting basic research in fields closely associated to their missions. The Foundation is not expected to have responsibility for the applied research and development programs of other agencies, and each agency is accountable for the scope and quality of its developmental effort.

With respect to the evaluation function, the Foundation has consistently pointed out that it is unrealistic to expect one agency to render judgment on the over-all performance of another agency unless an agency requests such help. The Foundation has chosen instead to approach the problem in terms of specific areas of science. Through close liaison and exchange of information with other science agencies, the Foundation has endeavored to identify areas that are receiving inadequate support or that require attention for other reasons. In this way it has been possible to bring about needed adjustments on an amicable, cooperative basis.

Executive Order 10807 of March 13, 1959, establishing the Federal Council on Science and Technology, also redefines the Foundation's role in the development of national science policy as applying only to basic research. Within this more specialized framework, the Foundation has been steadily formulating national science policies in the course of day-to-day operations, frequently on the basis of agreement and understanding with other agencies. Those who insist that policy must be handed down "ready made" in the form of a proclamation or edict do not understand the policy-making process. To be workable, policy must evolve on the basis of experience.

In 1959 the Foundation listed a compilation of some fifty science policies of a government-wide, national character that had been recommended by the Foundation during the previous eight years. Drawn from a variety of public statements and published reports the policies are grouped under the following broad categories: Basic Research; Government-University Relationships in the Conduct of Federally Sponsored Research; Indirect Costs; Education and Training; Federal Financial Support of Research Facilities; Government-Industry Relationships on Research; International Scientific Activities; Organization and Administration of Research; Medical Research, and Scientific Information.

As background data for its own research programs and for policy formulation concerning the role of the Federal Government in the support of science, the Foundation early established a series of studies of the nature and extent of the national effort in research and development. Comprehensive surveys are made on a recurring basis of the research and development effort of industry and of universities and other nonprofit institutions. The Foundation's analyses of the support of research and development by Federal agencies are published annually in *Federal Funds for Science*. In addition to statistical surveys of the volume of research and development, the Foundation is also engaged in analytical studies of the close relationship that exists between research and development and the economy in order to achieve a fuller understanding of the effects of research and development on various economic and industrial activities. The whole effort carries out the directive in the Executive Order "to make comprehensive studies and recommendations regarding the Nation's research effort and its resources for scientific activities. . . ."

A final word about the Foundation's budget is perhaps of interest. The following table summarizes Dr. Bush's projected budget for the National Research Foundation and the National Science Foundation's actual appropriation for its *fifth operating* year, fiscal year 1956; 1952 was the first year for which operating funds ($3.5 million) were appropriated for the Foundation by Congress.

(millions of dollars)

Activity [1]	Bush's Budget for the National Research Foundation		NSF funds [2] 5th Year 9th Year	
	1st Year	5th Year	(FY 1956)	(FY 1960)
Division of Medical Research	5.0	20.0	— [3]	— [3]
Division of Natural Sciences [3]	10.0	50.0	10.0	67.1
Division of National Defense	10.0	20.0	—	—
Division of Scientific Personnel and Education	7.0	29.0	3.4	64.5
Division of Publications & Scientific Collaboration [4]	.5	1.0	.4	5.4
Administration	1.0	2.5	1.3	6.2
Other [5]	—	—	.9	16.0
Total	33.5	122.5	16.0	159.2

[1] Except for the "other" category, these were the activities enumerated by Dr. Bush in his projected budget for the Foundation.

[2] Fiscal year 1956 data from The Budget of the United States, 1956, p. 159. In fiscal year 1960 the Foundation's total adjusted appropriation amounts to $154.8 million. The total of $159.2 million, shown here, includes $4.4 million carried forward from fiscal year 1959. See part 3, *Hearings before the Subcommittee on Appropriations, House of Representatives, Eighty-Sixth Congress, Second Session, Independent Offices Appropriations for 1961*, for further details on program activities.

[3] Funds administered by the Foundation's Biological and Medical Sciences Division are included in the total shown for the Division of Natural Sciences. This total also includes grants for the social sciences.

[4] Scientific information program activities are administered in the Foundation by the Office of Science Information Service.

[5] Includes funds for facilities, other program activities, and all adjustments.

A glance at the chart indicates that by the end of the fifth year the Foundation was operating at a level fifty per cent lower than that recommended by Dr. Bush for the first year. By the fifth year the National Science Foundation was operating at about thirteen per cent of the level suggested by Dr. Bush for that year. By 1960, however, the Foundation's appropriation for all activities was $159,200,000, almost ten times the 1956 level.

In order to understand the whole support situation it is necessary to look beyond a bare statistical comparison of Dr. Bush's recommendations and the Foundation's financial resources. As previously mentioned, Dr. Bush had visualized the Foundation as the sole support of basic research in the Government. This has been far from the fact. As already noted, a number of agencies began actively to support basic research during the five years of legislative debate of the National Science Foundation bills. It is estimated that in 1956 the Federal Government obligated about $200 million for basic research. Of this amount somewhat less than $120 million went for basic research related to "national defense" (Department of Defense $72 million, and Atomic Energy Commission $45 million). Twenty-six million dollars represents the total basic research reported by the National Institutes of Health for the year. The remainder of the $200 million is variously distributed among the Departments of Agriculture, Commerce, Interior, National Advisory Committee for Aeronautics, National Science Foundation, and the Smithsonian Institution.

Rough estimates indicate that about $115 million of the $200 million total 1956 obligation for basic research went to nonprofit institutions, including colleges and universities, research centers, research institutions, hospitals, and so on. Thus it would appear from these estimates that although the Foundation itself had not reached the projected level of basic research support proposed for its fifth year the Federal Government as a whole was providing the kind of basic research support visualized by Dr. Bush at a level somewhat higher than he projected.

In the history of the National Science Foundation's appropriations one is able to trace something of the public reaction to the international and national political situation. A $15 million limit on the Foundation's appropriations had been written into the law.* For its first year of operations, however, Congress appropriated to the Foundation only a small fraction of that amount —$3.5 million. Appropriations for the Foundation climbed slowly but steadily as Congress gained confidence in its operations and possibly also as a result of some dawning recognition on the part of the public of the importance of basic research. By the fifth year, 1956, the appropriation was up to $16 million. In the summer of 1955 the Foundation published a National Research Council study, *Soviet Professional Manpower*, which drew sobering comparisons between the rates at which the U. S. and the U.S.S.R. are training scientific and technical manpower. One result of these findings was that the Congress sharply increased Foundation funds for education in the sciences. The Foundation appropriation for fiscal year 1957, $40 million, more than doubled that of the preceding year. The next large increment came in 1959 when $130 million was appropriated in the wake of intense national concern over the Russian sputnik and all that it implied. Funds available for fiscal year 1960 total more than $159 million.

What can be said in summation? The principal mechanisms recommended by Dr. Bush for the support and encouragement of basic research and education in the sciences have been realized. All branches of the Federal Government have recognized the importance of these matters to the public welfare,

---

* This limitation was repealed by Act of Aug. 8, 1953 (67 Stat. 488).

and support is available in a variety of forms from a variety of sources. The universities, which have been the principal recipients of support, have expressed their approval of this diversity in the sources of support. Such diversity has meant more funds, greater flexibility, and the possibility of more than one approach.

It is difficult to say what the optimum level of support should be except to recognize that at some point a finite limit is set by the number of competent investigators available. At the present time the ratio of basic research funds to the over-all research and development funds of the Federal budget is something like seven per cent. Undoubtedly the ratio should be higher.

We have the organization; to a considerable extent we have the dollars, people, and facilities. Can we conclude, then, that the objectives of the Bush Report have been fully met? When one has been very close to the scene it is not possible to speak with complete objectivity and detachment. I think it can be said that the Government is doing well, both in the provision of funds and in the exercise of leadership. There remains, however, one conspicuous difficulty to be overcome. It is that people generally still do not clearly understand and appreciate the importance of education and the importance of science as distinguished from technology. As Dr. Bush so trenchantly observed:

The distinction between applied and pure research is not a hard and fast one, and industrial scientists may tackle specific problems from broad fundamental viewpoints. But it is important to emphasize that there is a perverse law governing research: under the pressure for immediate results, and unless deliberate policies are set up to guard against this, *applied research invariably drives out pure.*

This moral is clear: It is pure research which deserves and requires special protection and specially assured support.

It must be admitted that as a people and a Nation we have not been properly appreciative of intellectual achievement. This awareness and appreciation is not something the Government can legislate into being. We must build it into our national consciousness through our educational system, and until we do, science and all other forms of intellectual activity will lack the firm foundation they require.

LETTER OF TRANSMITTAL

OFFICE OF SCIENTIFIC RESEARCH AND DEVELOPMENT
1530 P Street, NW.
WASHINGTON 25, D. C.

JULY 5, 1945.

DEAR MR. PRESIDENT:

In a letter dated November 17, 1944, President Roosevelt requested my recommendations on the following points:

(1) What can be done, consistent with military security, and with the prior approval of the military authorities, to make known to the world as soon as possible the contributions which have been made during our war effort to scientific knowledge?

(2) With particular reference to the war of science against disease, what can be done now to organize a program for continuing in the future the work which has been done in medicine and related sciences?

(3) What can the Government do now and in the future to aid research activities by public and private organizations?

(4) Can an effective program be proposed for discovering and developing scientific talent in American youth so that the continuing future of scientific research in this country may be assured on a level comparable to what has been done during the war?

It is clear from President Roosevelt's letter that in speaking of science he had in mind the natural sciences, including biology and medicine, and I have so interpreted his questions. Progress in other fields, such as the social sciences and the humanities, is likewise important; but the program for science presented in my report warrants immediate attention.

In seeking answers to President Roosevelt's questions I have had the assistance of distinguished committees specially qualified to advise in respect to these subjects. The committees have given these matters the serious attention they deserve; indeed, they have regarded this as an opportunity to participate in shaping the policy of the country with reference to scientific research. They have had many meetings and have submitted formal reports. I have been in close touch with the work of the committees and with their members throughout. I have examined all of the data they assembled and the suggestions they submitted on the points raised in President Roosevelt's letter.

1

Although the report which I submit herewith is my own, the facts, conclusions, and recommendations are based on the findings of the committees which have studied these questions. Since my report is necessarily brief, I am including as appendices the full reports of the committees.

A single mechanism for implementing the recommendations of the several committees is essential. In proposing such a mechanism I have departed somewhat from the specific recommendations of the committees, but I have since been assured that the plan I am proposing is fully acceptable to the committee members.

The pioneer spirit is still vigorous within this Nation. Science offers a largely unexplored hinterland for the pioneer who has the tools for his task. The rewards of such exploration both for the Nation and the individual are great. Scientific progress is one essential key to our security as a nation, to our better health, to more jobs, to a higher standard of living, and to our cultural progress.

Respectfully yours,

(s)   V. Bush, Director.

THE PRESIDENT OF THE UNITED STATES,
The White House,
Washington, D. C.

2

PRESIDENT ROOSEVELT'S LETTER

THE WHITE HOUSE
Washington, D. C.
November 17, 1944

DEAR DR. BUSH:

The Office of Scientific Research and Development, of which you are the Director, represents a unique experiment of team-work and cooperation in coordinating scientific research and in applying existing scientific knowledge to the solution of the technical problems paramount in war. Its work has been conducted in the utmost secrecy and carried on without public recognition of any kind; but its tangible results can be found in the communiques coming in from the battlefronts all over the world. Some day the full story of its achievements can be told.

There is, however, no reason why the lessons to be found in this experiment cannot be profitably employed in times of peace. The information, the techniques, and the research experience developed by the Office of Scientific Research and Development and by the thousands of scientists in the universities and in private industry, should be used in the days of peace ahead for the improvement of the national health, the creation of new enterprises bringing new jobs, and the betterment of the national standard of living.

It is with that objective in mind that I would like to have your recommendations on the following four major points:

First: What can be done, consistent with military security, and with the prior approval of the military authorities, to make known to the world as soon as possible the contributions which have been made during our war effort to scientific knowledge?

The diffusion of such knowledge should help us stimulate new enterprises, provide jobs for our returning servicemen and other workers, and make possible great strides for the improvement of the national well-being.

Second: With particular reference to the war of science against disease, what can be done now to organize a program for continuing in the future the work which has been done in medicine and related sciences?

The fact that the annual deaths in this country from one or two diseases alone are far in excess of the total number of lives lost by us in battle during this war should make us conscious of the duty we owe future generations.

Third: What can the Government do now and in the future to aid research activities by public and private organizations? The proper roles of public and of private research, and their interrelation, should be carefully considered.

3

Fourth: Can an effective program be proposed for discovering and developing scientific talent in American youth so that the continuing future of scientific research in this country may be assured on a level comparable to what has been done during the war?

New frontiers of the mind are before us, and if they are pioneered with the same vision, boldness, and drive with which we have waged this war we can create a fuller and more fruitful employment and a fuller and more fruitful life.

I hope that, after such consultation as you may deem advisable with your associates and others, you can let me have your considered judgment on these matters as soon as convenient—reporting on each when you are ready, rather than waiting for completion of your studies in all.

Very sincerely yours,

(s) FRANKLIN D. ROOSEVELT.

DR. VANNEVAR BUSH,
Office of Scientific Research and Development,
Washington, D. C.

SUMMARY OF THE REPORT

Scientific Progress Is Essential

Progress in the war against disease depends upon a flow of new scientific knowledge. New products, new industries, and more jobs require continuous additions to knowledge of the laws of nature, and the application of that knowledge to practical purposes. Similarly, our defense against aggression demands new knowledge so that we can develop new and improved weapons. This essential, new knowledge can be obtained only through basic scientific research.

Science can be effective in the national welfare only as a member of a team, whether the conditions be peace or war. But without scientific progress no amount of achievement in other directions can insure our health, prosperity, and security as a nation in the modern world.

For the War Against Disease

We have taken great strides in the war against disease. The death rate for all diseases in the Army, including overseas forces, has been reduced from 14.1 per thousand in the last war to 0.6 per thousand in this war. In the last 40 years life expectancy has increased from 49 to 65 years, largely as a consequence of the reduction in the death rates of infants and children. But we are far from the goal. The annual deaths from one or two diseases far exceed the total number of American lives lost in battle during this war. A large fraction of these deaths in our civilian population cut short the useful lives of our citizens. Approximately 7,000,000 persons in the United States are mentally ill and their care costs the public over $175,000,000 a year. Clearly much illness remains for which adequate means of prevention and cure are not yet known.

The responsibility for basic research in medicine and the underlying sciences, so essential to progress in the war against disease, falls primarily upon the medical schools and universities. Yet we find that the traditional sources of support for medical research in the medical schools and universities, largely endowment income, foundation grants, and private donations, are diminishing and there is no immediate prospect of a change in this trend. Meanwhile, the cost of medical research has been rising. If we are to maintain the progress in medicine which has marked the last 25 years, the Government should extend financial support to basic medical research in the medical schools and in universities.

For Our National Security

The bitter and dangerous battle against the U-boat was a battle of scientific techniques—and our margin of success was dangerously small. The new eyes which radar has supplied can sometimes be blinded by new scientific developments. V–2 was countered only by capture of the launching sites.

We cannot again rely on our allies to hold off the enemy while we struggle to catch up. There must be more—and more adequate—military research in peacetime. It is essential that the civilian scientists continue in peacetime some portion of those contributions to national security which they have made so effectively during the war. This can best be done through a civilian-controlled organization with close liaison with the Army and Navy, but with funds direct from Congress, and the clear power to initiate military research which will supplement and strengthen that carried on directly under the control of the Army and Navy.

And for the Public Welfare

One of our hopes is that after the war there will be full employment. To reach that goal the full creative and productive energies of the American people must be released. To create more jobs we must make new and better and cheaper products. We want plenty of new, vigorous enterprises. But new products and processes are not born full-grown. They are founded on new principles and new conceptions which in turn result from basic scientific research. Basic scientific research is scientific capital. Moreover, we cannot any longer depend upon Europe as a major source of this scientific capital. Clearly, more and better scientific research is one essential to the achievement of our goal of full employment.

How do we increase this scientific capital? First, we must have plenty of men and women trained in science, for upon them depends both the creation of new knowledge and its application to practical purposes. Second, we must strengthen the centers of basic research which are principally the colleges, universities, and research institutes. These institutions provide the environment which is most conducive to the creation of new scientific knowledge and least under pressure for immediate, tangible results. With some notable exceptions, most research in industry and in Government involves application of existing scientific knowledge to practical problems. It is only the colleges, universities, and a few research institutes that devote most of their research efforts to expanding the frontiers of knowledge.

Expenditures for scientific research by industry and Government increased from $140,000,000 in 1930 to $309,000,000 in 1940. Those for the colleges and universities increased from $20,000,000 to $31,000,000, while those for research institutes declined from $5,200,000 to $4,500,000 during the same period. If the colleges, universities, and research institutes are to meet the rapidly increasing demands of industry and Government for new scientific knowledge, their basic research should be strengthened by use of public funds.

For science to serve as a powerful factor in our national welfare, applied research both in Government and in industry must be vigorous. To improve the quality of scientific research within the Government, steps should be taken to modify the procedures for recruiting, classifying, and compensating scientific personnel in order to reduce the present handicap of governmental scientific bureaus in competing with industry and the universities for top-grade scientific talent. To provide coordination of the common scientific activities of these governmental agencies as to policies and budgets, a permanent Science Advisory Board should be created to advise the executive and legislative branches of Government on these matters.

The most important ways in which the Government can promote industrial research are to increase the flow of new scientific knowledge through support of basic research, and to aid in the development of scientific talent. In addition, the Government should provide suitable incentives to industry to conduct research (a) by clarification of present uncertainties in the Internal Revenue Code in regard to the deductibility of research and development expenditures as current charges against net income, and (b) by strengthening the patent system so as to eliminate uncertainties which now bear heavily on small industries and so as to prevent abuses which reflect discredit upon a basically sound system. In addition, ways should be found to cause the benefits of basic research to reach industries which do not now utilize new scientific knowledge.

We Must Renew Our Scientific Talent

The responsibility for the creation of new scientific knowledge—and for most of its application—rests on that small body of men and women who understand the fundamental laws of nature and are skilled in the techniques of scientific research. We shall have rapid or slow advance on any scientific frontier depending on the number of highly qualified and trained scientists exploring it.

The deficit of science and technology students who, but for the war, would have received bachelor's degrees is about 150,000. It is estimated that the deficit of those obtaining advanced degrees in these fields will amount in 1955 to about 17,000—for it takes at least 6 years from college entry to achieve a doctor's degree or its equivalent in science or engineering. The real ceiling on our productivity of new scientific knowledge and its application in the war against disease, and the development of new products and new industries, is the number of trained scientists available.

The training of a scientist is a long and expensive process. Studies clearly show that there are talented individuals in every part of the population, but with few exceptions, those without the means of buying higher education

7

go without it. If ability, and not the circumstance of family fortune, determines who shall receive higher education in science, then we shall be assured of constantly improving quality at every level of scientific activity. The Government should provide a reasonable number of undergraduate scholarships and graduate fellowships in order to develop scientific talent in American youth. The plans should be designed to attract into science only that proportion of youthful talent appropriate to the needs of science in relation to the other needs of the Nation for high abilities.

Including Those in Uniform

The most immediate prospect of making up the deficit in scientific personnel is to develop the scientific talent in the generation now in uniform. Even if we should start now to train the current crop of high-school graduates none would complete graduate studies before 1951. The Armed Services should comb their records for men who, prior to or during the war, have given evidence of talent for science, and make prompt arrangements, consistent with current discharge plans, for ordering those who remain in uniform, as soon as militarily possible, to duty at institutions here and overseas where they can continue their scientific education. Moreover, the Services should see that those who study overseas have the benefit of the latest scientific information resulting from research during the war.

The Lid Must Be Lifted

While most of the war research has involved the application of existing scientific knowledge to the problems of war, rather than basic research, there has been accumulated a vast amount of information relating to the application of science to particular problems. Much of this can be used by industry. It is also needed for teaching in the colleges and universities here and in the Armed Forces Institutes overseas. Some of this information must remain secret, but most of it should be made public as soon as there is ground for belief that the enemy will not be able to turn it against us in this war. To select that portion which should be made public, to coordinate its release, and definitely to encourage its publication, a Board composed of Army, Navy, and civilian scientific members should be promptly established.

A Program for Action

The Government should accept new responsibilities for promoting the flow of new scientific knowledge and the development of scientific talent in our youth. These responsibilities are the proper concern of the Government, for they vitally affect our health, our jobs, and our national security. It is in keeping also with basic United States policy that the Government should foster the opening of new frontiers and this is the modern way to do it. For many years the Government has wisely supported research in the agricultural colleges and the benefits have been great. The time has come when such support should be extended to other fields.

The effective discharge of these new responsibilities will require the full attention of some over-all agency devoted to that purpose. There is not now in the permanent governmental structure receiving its funds from Congress an agency adapted to supplementing the support of basic research in the colleges, universities, and research institutes, both in medicine and the natural sciences, adapted to supporting research on new weapons for both Services, or adapted to administering a program of science scholarships and fellowships.

Therefore I recommend that a new agency for these purposes be established. Such an agency should be composed of persons of broad interest and experience, having an understanding of the peculiarities of scientific research and scientific education. It should have stability of funds so that long-range programs may be undertaken. It should recognize that freedom of inquiry must be preserved and should leave internal control of policy, personnel, and the method and scope of research to the institutions in which it is carried on. It should be fully responsible to the President and through him to the Congress for its program.

Early action on these recommendations is imperative if this Nation is to meet the challenge of science in the crucial years ahead. On the wisdom with which we bring science to bear in the war against disease, in the creation of new industries, and in the strengthening of our Armed Forces depends in large measure our future as a nation.

Part One

INTRODUCTION

Scientific Progress Is Essential

We all know how much the new drug, penicillin, has meant to our grievously wounded men on the grim battlefronts of this war—the countless lives it has saved—the incalculable suffering which its use has prevented. Science and the great practical genius of this Nation made this achievement possible.

Some of us know the vital role which radar has played in bringing the Allied Nations to victory over Nazi Germany and in driving the Japanese steadily back from their island bastions. Again it was painstaking scientific research over many years that made radar possible.

What we often forget are the millions of pay envelopes on a peacetime Saturday night which are filled because new products and new industries have provided jobs for countless Americans. Science made that possible, too.

In 1939 millions of people were employed in industries which did not even exist at the close of the last war—radio, air conditioning, rayon and other synthetic fibers, and plastics are examples of the products of these industries. But these things do not mark the end of progress—they are but the beginning if we make full use of our scientific resources. New manufacturing industries can be started and many older industries greatly strengthened and expanded if we continue to study nature's laws and apply new knowledge to practical purposes.

Great advances in agriculture are also based upon scientific research. Plants which are more resistant to disease and are adapted to short growing seasons, the prevention and cure of livestock diseases, the control of our insect enemies, better fertilizers, and improved agricultural practices, all stem from painstaking scientific research.

Advances in science when put to practical use mean more jobs, higher wages, shorter hours, more abundant crops, more leisure for recreation, for study, for learning how to live without the deadening drudgery which has been the burden of the common man for ages past. Advances in science will also bring higher standards of living, will lead to the prevention or cure of diseases, will promote conservation of our limited national resources, and will assure means of defense against aggression. But to achieve these objectives— to secure a high level of employment, to maintain a position of world leadership—the flow of new scientific knowledge must be both continuous and substantial.

Our population increased from 75 million to 130 million between 1900 and 1940. In some countries comparable increases have been accompanied by famine. In this country the increase has been accompanied by more abundant food supply, better living, more leisure, longer life, and better health. This is, largely, the product of three factors—the free play of initiative of a vigorous people under democracy, the heritage of great natural wealth, and the advance of science and its application.

Science, by itself, provides no panacea for individual, social, and economic ills. It can be effective in the national welfare only as a member of a team, whether the conditions be peace or war. But without scientific progress no amount of achievement in other directions can insure our health, prosperity, and security as a nation in the modern world.

Science Is a Proper Concern of Government

It has been basic United States policy that Government should foster the opening of new frontiers. It opened the seas to clipper ships and furnished land for pioneers. Although these frontiers have more or less disappeared, the frontier of science remains. It is in keeping with the American tradition —one which has made the United States great—that new frontiers shall be made accessible for development by all American citizens.

Moreover, since health, well-being, and security are proper concerns of Government, scientific progress is, and must be, of vital interest to Government. Without scientific progress the national health would deteriorate; without scientific progress we could not hope for improvement in our standard of living or for an increased number of jobs for our citizens; and without scientific progress we could not have maintained our liberties against tyranny.

Government Relations to Science—Past and Future

From early days the Government has taken an active interest in scientific matters. During the nineteenth century the Coast and Geodetic Survey, the Naval Observatory, the Department of Agriculture, and the Geological Survey were established. Through the Land Grant College Acts the Government has supported research in state institutions for more than 80 years on a gradually increasing scale. Since 1900 a large number of scientific agencies have been established within the Federal Government, until in 1939 they numbered more than 40.

Much of the scientific research done by Government agencies is intermediate in character between the two types of work commonly referred to as basic and applied research. Almost all Government scientific work has ultimate practical objectives but, in many fields of broad national concern, it commonly involves long-term investigation of a fundamental nature. Generally speaking, the scientific agencies of Government are not so concerned with immediate practical objectives as are the laboratories of industry nor, on the other hand, are they as free to explore any natural phenomena without regard to possible economic applications as are the educational and private research institutions. Government scientific agencies have splendid records of achievement, but they are limited in function.

11

We have no national policy for science. The Government has only begun to utilize science in the Nation's welfare. There is no body within the Government charged with formulating or executing a national science policy. There are no standing committees of the Congress devoted to this important subject. Science has been in the wings. It should be brought to the center of the stage—for in it lies much of our hope for the future.

There are areas of science in which the public interest is acute but which are likely to be cultivated inadequately if left without more support than will come from private sources. These areas—such as research on military problems, agriculture, housing, public health, certain medical. research, and research involving expensive capital facilities beyond the capacity of private institutions—should be advanced by active Government support. To date, with the exception of the intensive war research conducted by the Office of Scientific Research and Development, such support has been meager and intermittent.

For reasons presented in this report we are entering a period when science needs and deserves increased support from public funds.

Freedom of Inquiry Must Be Preserved

The publicly and privately supported colleges, universities, and research institutes are the centers of basic research. They are the wellsprings of knowledge and understanding. As long as they are vigorous and healthy and their scientists are free to pursue the truth wherever it may lead, there will be a flow of new scientific knowledge to those who can apply it to practical problems in Government, in industry, or elsewhere.

Many of the lessons learned in the war-time application of science under Government can be profitably applied in peace. The Government is peculiarly fitted to perform certain functions, such as the coordination and support of broad programs on problems of great national importance. But we must proceed with caution in carrying over the methods which work in wartime to the very different conditions of peace. We must remove the rigid controls which we have had to impose, and recover freedom of inquiry and that healthy competitive scientific spirit so necessary for expansion of the frontiers of scientific knowledge.

Scientific progress on a broad front results from the free play of free intellects, working on subjects of their own choice, in the manner dictated by their curiosity for exploration of the unknown. Freedom of inquiry must be preserved under any plan for Government support of science in accordance with the Five Fundamentals listed on page 32.

The study of the momentous questions presented in President Roosevelt's letter has been made by able committees working diligently. This report presents conclusions and recommendations based upon the studies of these committees which appear in full as the appendices. Only in the creation of one over-all mechanism rather than several does this report depart from the specific recommendations of the committees. The members of the committees have reviewed the recommendations in regard to the single mechanism and have found this plan thoroughly acceptable.

Part Two

THE WAR AGAINST DISEASE

In War

The death rate for all diseases in the Army, including the overseas forces, has been reduced from 14.1 per thousand in the last war to 0.6 per thousand in this war.

Such ravaging diseases as yellow fever, dysentery, typhus, tetanus, pneumonia, and meningitis have been all but conquered by penicillin and the sulfa drugs, the insecticide DDT, better vaccines, and improved hygienic measures. Malaria has been controlled. There has been dramatic progress in surgery.

The striking advances in medicine during the war have been possible only because we had a large backlog of scientific data accumulated through basic research in many scientific fields in the years before the war.

In Peace

In the last 40 years life expectancy in the United States has increased from 49 to 65 years largely as a consequence of the reduction in the death rates of infants and children; in the last 20 years the death rate from the diseases of childhood has been reduced 87 percent.

Diabetes has been brought under control by insulin, pernicious anemia by liver extracts; and. the once widespread deficiency diseases have been much reduced, even in the lowest income groups, by accessory food factors and improvement of diet. Notable advances have been made in the early diagnosis of cancer, and in the surgical and radiation treatment of the disease.

These results have been achieved through a great amount of basic research in medicine and the preclinical sciences, and by the dissemination of this new scientific knowledge through the physicians and medical services and public health agencies of the country. In this cooperative endeavor the pharmaceutical industry has played an important role, especially during the war. All of the medical and public health groups share credit for these achievements; they form interdependent members of a team.

Progress in combating disease depends upon an expanding body of new scientific knowledge.

Unsolved Problems

As President Roosevelt observed, the annual deaths from one or two diseases are far in excess of the total number of American lives lost in battle during this war. A large fraction of these deaths in our civilian population cut short the useful lives of our citizens. This is our present position despite the fact that in the last three decades notable progress has been made in civilian medicine. The reduction in death rate from diseases of childhood has shifted the emphasis to the middle and old age groups, particularly to the malignant diseases and the degenerative processes prominent in later life. Cardiovascular disease, including chronic disease of the kidneys, arteriosclerosis, and cerebral hemorrhage, now account for 45 percent of the deaths in the United States. Second are the infectious diseases, and third is cancer. Added to these are many maladies (for example, the common cold, arthritis, asthma and hay fever, peptic ulcer) which, though infrequently fatal, cause incalculable disability.

Another aspect of the changing emphasis is the increase of mental diseases. Approximately 7 million persons in the United States are mentally ill; more than one-third of the hospital beds are occupied by such persons, at a cost of $175 million a year. Each year 125,000 new mental cases are hospitalized.

Notwithstanding great progress in prolonging the span of life and in relief of suffering, much illness remains for which adequate means of prevention and cure are not yet known. While additional physicians, hospitals, and health programs are needed, their full usefulness cannot be attained unless we enlarge our knowledge of the human organism and the nature of disease. Any extension of medical facilities must be accompanied by an expanded program of medical training and research.

Broad and Basic Studies Needed

Discoveries pertinent to medical progress have often come from remote and unexpected sources, and it is certain that this will be true in the future. It is wholly probable that progress in the treatment of cardiovascular disease, renal disease, cancer, and similar refractory diseases will be made as the result of fundamental discoveries in subjects unrelated to those diseases, and perhaps entirely unexpected by the investigator. Further progress requires that the entire front of medicine and the underlying sciences of chemistry, physics, anatomy, biochemistry, physiology, pharmacology, bacteriology, pathology, parasitology, etc., be broadly developed.

Progress in the war against disease results from discoveries in remote and unexpected fields of medicine and the underlying sciences.

Coordinated Attack on Special Problems

Penicillin reached our troops in time to save countless lives because the Government coordinated and supported the program of research and development on the drug. The development moved from the early laboratory stage to large scale production and use in a fraction of the time it would have taken without such leadership. The search for better anti-malarials, which proceeded at a moderate tempo for many years, has been accelerated enormously by Government support during the war. Other examples can be cited in which medical progress has been similarly advanced. In achieving these results, the Government has provided over-all coordination and support; it has not dictated how the work should be done within any cooperating institution.

Discovery of new therapeutic agents and methods usually results from basic studies in medicine and the underlying sciences. The development of such materials and methods to the point at which they become available to medical practitioners requires teamwork involving the medical schools, the science departments of universities, Government and the pharmaceutical industry. Government initiative, support, and coordination can be very effective in this development phase.

Government initiative and support for the development of newly discovered therapeutic materials and methods can reduce the time required to bring the benefits to the public.

Action is Necessary

The primary place for medical research is in the medical schools and universities. In some cases coordinated direct attack on special problems may be made by teams of investigators, supplementing similar attacks carried on by the Army, Navy, Public Health Service, and other organizations. Apart from teaching, however, the primary obligation of the medical schools and universities is to continue the traditional function of such institutions, namely, to provide the individual worker with an opportunity for free, untrammeled study of nature, in the directions and by the methods suggested by his interests, curiosity, and imagination. The history of medical science teaches clearly the supreme importance of affording the prepared mind complete freedom for the exercise of initiative. It is the special province of the medical schools and universities to foster medical research in this way—a duty which cannot be shifted to Government agencies, industrial organizations, or to any other institutions.

Where clinical investigations of the human body are required, the medical schools are in a unique position, because of their close relationship to teaching hospitals, to integrate such investigations with the work of the departments of preclinical science, and to impart new knowledge to physicians in training. At the same time, the teaching hospitals are especially well qualified to carry on medical research because of their close connection with the medical schools, on which they depend for staff and supervision.

Between World War I and World War II the United States overtook all other nations in medical research and assumed a position of world leadership. To a considerable extent this progress reflected the liberal financial support from university endowment income, gifts from individuals, and foundation grants in the 20's. The growth of research departments in medical schools has been very uneven, however, and in consequence most of the important work has been done in a few large schools. This should be corrected by building up the weaker institutions, especially in regions which now have no strong medical research activities.

The traditional sources of support for medical research, largely endowment income, foundation grants, and private donations, are diminishing, and there is no immediate prospect of a change in this trend. Meanwhile, research costs have steadily risen. More elaborate and expensive equipment is required, supplies are more costly, and the wages of assistants are higher. Industry is only to a limited extent a source of funds for basic medical research.

It is clear that if we are to maintain the progress in medicine which has marked the last 25 years, the Government should extend financial support to basic medical research in the medical schools and in the universities, through grants both for research and for fellowships. The amount which can be effectively spent in the first year should not exceed 5 million dollars. After a program is under way perhaps 20 million dollars a year can be spent effectively.

Part Three

SCIENCE AND THE PUBLIC WELFARE

Relation to National Security

In this war it has become clear beyond all doubt that scientific research is absolutely essential to national security. The bitter and dangerous battle against the U-boat was a battle of scientific techniques—and our margin of success was dangerously small. The new eyes which radar supplied to our fighting forces quickly evoked the development of scientific countermeasures which could often blind them. This again represents the ever continuing battle of techniques. The V–1 attack on London was finally defeated by three devices developed during this war and used superbly in the field. V–2 was countered only by capture of the launching sites.

The Secretaries of War and Navy recently stated in a joint letter to the National Academy of Sciences:

This war emphasizes three facts of supreme importance to national security: (1) Powerful new tactics of defense and offense are developed around new weapons created by scientific and engineering research; (2) the competitive time element in developing those weapons and tactics may be decisive; (3) war is increasingly total war, in which the armed services must be supplemented by active participation of every element of civilian population.

To insure continued preparedness along farsighted technical lines, the research scientists of the country must be called upon to continue in peacetime some substantial portion of those types of contribution to national security which they have made so effectively during the stress of the present war * * *.

There must be more—and more adequate—military research during peacetime. We cannot again rely on our allies to hold off the enemy while we struggle to catch up. Further, it is clear that only the Government can undertake military research; for it must be carried on in secret, much of it has no commercial value, and it is expensive. The obligation of Government to support research on military problems is inescapable.

Modern war requires the use of the most advanced scientific techniques. Many of the leaders in the development of radar are scientists who before the war had been exploring the nucleus of the atom. While there must be increased emphasis on science in the future training of officers for both the Army and Navy, such men cannot be expected to be specialists in scientific research. Therefore, a professional partnership between the officers in the Services and civilian scientists is needed.

The Army and Navy should continue to carry on research and development on the improvement of current weapons. For many years the National Advisory Committee for Aeronautics has supplemented the work of the Army and Navy by conducting basic research on the problems of flight. There should now be permanent civilian activity to supplement the research work of the Services in other scientific fields so as to carry on in time of peace some part of the activities of the emergency wartime Office of Scientific Research and Development.

Military preparedness requires a permanent independent, civilian-controlled organization, having close liaison with the Army and Navy, but with funds directly from Congress and with the clear power to initiate military research which will supplement and strengthen that carried on directly under the control of the Army and Navy.

Science and Jobs

One of our hopes is that after the war there will be full employment, and that the production of goods and services will serve to raise our standard of living. We do not know yet how we shall reach that goal, but it is certain that it can be achieved only by releasing the full creative and productive energies of the American people.

Surely we will not get there by standing still, merely by making the same things we made before and selling them at the same or higher prices. We will not get ahead in international trade unless we offer new and more attractive and cheaper products.

Where will these new products come from? How will we find ways to make better products at lower cost? The answer is clear. There must be a stream of new scientific knowledge to turn the wheels of private and public enterprise. There must be plenty of men and women trained in science and technology for upon them depend both the creation of new knowledge and its application to practical purposes.

More and better scientific research is essential to the achievement of our goal of full employment.

The Importance of Basic Research

Basic research is performed without thought of practical ends. It results in general knowledge and an understanding of nature and its laws. This general knowledge provides the means of answering a large number of important practical problems, though it may not give a complete specific answer to any one of them. The function of applied research is to provide such complete answers. The scientist doing basic research may not be at all interested in the practical applications of his work, yet the further progress of industrial development would eventually stagnate if basic scientific research were long neglected.

One of the peculiarities of basic science is the variety of paths which lead to productive advance. Many of the most important discoveries have come as a result of experiments undertaken with very different purposes in mind. Statistically it is certain that important and highly useful discoveries will result from some fraction of the undertakings in basic science; but the results of any one particular investigation cannot be predicted with accuracy.

Basic research leads to new knowledge. It provides scientific capital. It creates the fund from which the practical applications of knowledge must be drawn. New products and new processes do not appear full-grown. They are founded on new principles and new conceptions, which in turn are painstakingly developed by research in the purest realms of science.

Today, it is truer than ever that basic research is the pacemaker of technological progress. In the nineteenth century, Yankee mechanical ingenuity, building largely upon the basic discoveries of European scientists, could greatly advance the technical arts. Now the situation is different.

A nation which depends upon others for its new basic scientific knowledge will be slow in its industrial progress and weak in its competitive position in world trade, regardless of its mechanical skill.

Centers of Basic Research

Publicly and privately supported colleges and universities and the endowed research institutes must furnish both the new scientific knowledge and the trained research workers. These institutions are uniquely qualified by tradition and by their special characteristics to carry on basic research. They are charged with the responsibility of conserving the knowledge accumulated by the past, imparting that knowledge to students, and contributing new knowledge of all kinds. It is chiefly in these institutions that scientists may work in an atmosphere which is relatively free from the adverse pressure of convention, prejudice, or commercial necessity. At their best they provide the scientific worker with a strong sense of solidarity and security, as well as a substantial degree of personal intellectual freedom. All of these factors are of great importance in the development of new knowledge, since much of new knowledge is certain to arouse opposition because of its tendency to challenge current beliefs or practice.

Industry is generally inhibited by preconceived goals, by its own clearly defined standards, and by the constant pressure of commercial necessity. Satisfactory progress in basic science seldom occurs under conditions prevailing in the normal industrial laboratory. There are some notable exceptions, it is true, but even in such cases it is rarely possible to match the universities in respect to the freedom which is so important to scientific discovery.

To serve effectively as the centers of basic research these institutions must be strong and healthy. They must attract our best scientists as teachers and investigators. They must offer research opportunities and sufficient compensation to enable them to compete with industry and government for the cream of scientific talent.

During the past 25 years there has been a great increase in industrial research involving the application of scientific knowledge to a multitude of practical purposes—thus providing new products, new industries, new investment opportunities, and millions of jobs. During the same period research within Government—again largely applied research—has also been greatly expanded. In the decade from 1930 to 1940 expenditures for industrial research increased from $116,000,000 to $240,000,000 and those for scientific research in Government rose from $24,000,000 to $69,000,000. During the same period expenditures for scientific research in the colleges and universities increased from $20,000,000 to $31,000,000, while those in the endowed research institutes declined from $5,200,000 to $4,500,000. These are the best estimates available. The figures have been taken from a variety of sources and arbitrary definitions have necessarily been applied, but it is believed that they may be accepted as indicating the following trends:

(a) Expenditures for scientific research by industry and Government—almost entirely applied research—have more than doubled between 1930 and 1940. Whereas in 1930 they were six times as large as the research expenditures of the colleges, universities, and research institutes, by 1940 they were nearly ten times as large.

(b) While expenditures for scientific research in the colleges and universities increased by one-half during this period, those for the endowed research institutes have slowly declined.

If the colleges, universities, and research institutes are to meet the rapidly increasing demands of industry and Government for new scientific knowledge, their basic research should be strengthened by use of public funds.

Research Within the Government

Although there are some notable exceptions, most research conducted within governmental laboratories is of an applied nature. This has always been true and is likely to remain so. Hence Government, like industry, is dependent upon the colleges, universities, and research institutes to expand the basic scientific frontiers and to furnish trained scientific investigators.

Research within the Government represents an important part of our total research activity and needs to be strengthened and expanded after the war. Such expansion should be directed to fields of inquiry and service which are of public importance and are not adequately carried on by private organizations.

The most important single factor in scientific and technical work is the quality of personnel employed. The procedures currently followed within the Government for recruiting, classifying and compensating such personnel place the Government under a severe handicap in competing with industry and the universities for first-class scientific talent. Steps should be taken to reduce that handicap.

In the Government the arrangement whereby the numerous scientific agencies form parts of large departments has both advantages and disadvantages. But the present pattern is firmly established and there is much to be said for it. There is, however, a very real need for some measure of coordination of the common scientific activities of these agencies, both as to policies and budgets, and at present no such means exist.

A permanent Science Advisory Board should be created to consult with these scientific bureaus and to advise the executive and legislative branches of Government as to the policies and budgets of Government agencies engaged in scientific research.

This board should be composed of disinterested scientists who have no connection with the affairs of any Government agency.

Industrial Research

The simplest and most effective way in which the Government can strengthen industrial research is to support basic research and to develop scientific talent.

The benefits of basic research do not reach all industries equally or at the same speed. Some small enterprises never receive any of the benefits. It has been suggested that the benefits might be better utilized if "research clinics" for such enterprises were to be established. Businessmen would thus be able to make more use of research than they now do. This proposal is certainly worthy of further study.

One of the most important factors affecting the amount of industrial research is the income-tax law. Government action in respect to this subject will affect the rate of technical progress in industry. Uncertainties as to the attitude of the Bureau of Internal Revenue regarding the deduction of research and development expenses are a deterrent to research expenditure. These uncertainties arise from lack of clarity of the tax law as to the proper treatment of such costs.

The Internal Revenue Code should be amended to remove present uncertainties in regard to the deductibility of research and development expenditures as current charges against net income.

Research is also affected by the patent laws. They stimulate new invention and they make it possible for new industries to be built around new devices or new processes. These industries generate new jobs and new products, all of which contribute to the welfare and the strength of the country.

Yet, uncertainties in the operation of the patent laws have impaired the ability of small industries to translate new ideas into processes and products of value to the Nation. These uncertainties are, in part, attributable to the difficulties and expense incident to the operation of the patent system as it presently exists. These uncertainties are also attributable to the existence of certain abuses which have appeared in the use of patents. The abuses should be corrected. They have led to extravagantly critical attacks which tend to discredit a basically sound system.

It is important that the patent system continue to serve the country in the manner intended by the Constitution, for it has been a vital element in the industrial vigor which has distinguished this Nation.

The National Patent Planning Commission has reported on this subject. In addition, a detailed study, with recommendations concerning the extent to which modifications should be made in our patent laws is currently being made under the leadership of the Secretary of Commerce. It is recommended, therefore, that specific action with regard to the patent laws be withheld pending the submission of the report devoted exclusively to that subject.

International Exchange of Scientific Information

International exchange of scientific information is of growing importance. Increasing specialization of science will make it more important than ever that scientists in this country keep continually abreast of developments abroad. In addition, a flow of scientific information constitutes one facet of general international accord which should be cultivated.

The Government can accomplish significant results in several ways: by aiding in the arrangement of international science congresses, in the official accrediting of American scientists to such gatherings, in the official reception of foreign scientists of standing in this country, in making possible a rapid flow of technical information, including translation service, and possibly in the provision of international fellowships. Private foundations and other groups partially fulfill some of these functions at present, but their scope is incomplete and inadequate.

The Government should take an active role in promoting the international flow of scientific information.

The Special Need for Federal Support

We can no longer count on ravaged Europe as a source of fundamental knowledge. In the past we have devoted much of our best efforts to the application of such knowledge which has been discovered abroad. In the future we must pay increased attention to discovering this knowledge for ourselves particularly since the scientific applications of the future will be more than ever dependent upon such basic knowledge.

New impetus must be given to research in our country. Such new impetus can come promptly only from the Government. Expenditures for research in the colleges, universities, and research institutes will otherwise not be able to meet the additional demands of increased public need for research.

Further, we cannot expect industry adequately to fill the gap. Industry will fully rise to the challenge of applying new knowledge to new products. The commercial incentive can be relied upon for that. But basic research is essentially noncommercial in nature. It will not receive the attention it requires if left to industry.

For many years the Government has wisely supported research in the agricultural colleges and the benefits have been great. The time has come when such support should be extended to other fields.

In providing Government support, however, we must endeavor to preserve as far as possible the private support of research both in industry and in the colleges, universities, and research institutes. These private sources should continue to carry their share of the financial burden.

The Cost of a Program

It is estimated that an adequate program for Federal support of basic research in the colleges, universities, and research institutes and for financing important applied research in the public interest, will cost about 10 million dollars at the outset and may rise to about 50 million dollars annually when fully underway at the end of perhaps 5 years.

Part Four

RENEWAL OF OUR SCIENTIFIC TALENT

Nature of the Problem

The responsibility for the creation of new scientific knowledge rests on that small body of men and women who understand the fundamental laws of nature and are skilled in the techniques of scientific research. While there will always be the rare individual who will rise to the top without benefit of formal education and training, he is the exception and even he might make a more notable contribution if he had the benefit of the best education we have to offer. I cannot improve on President Conant's statement that:

"* * * in every section of the entire area where the word science may properly be applied, the limiting factor is a human one. We shall have rapid or slow advance in this direction or in that depending on the number of really first-class men who are engaged in the work in question. * * * So in the last analysis, the future of science in this country will be determined by our basic educational policy."

A Note of Warning

It would be folly to set up a program under which research in the natural sciences and medicine was expanded at the cost of the social sciences, humanities, and other studies so essential to national well-being. This point has been well stated by the Moe Committee as follows:

"As citizens, as good citizens, we therefore think that we must have in mind while examining the question before us—the discovery and development of scientific talent— the needs of the whole national welfare. We could not suggest to you a program which would syphon into science and technology a disproportionately large share of the Nation's highest abilities, without doing harm to the Nation, nor, indeed, without crippling science. * * * Science cannot live by and unto itself alone."

*   *   *   *   *   *   *   *

"The uses to which high ability in youth can be put are various and, to a large extent, are determined by social pressures and rewards. When aided by selective devices for picking out scientifically talented youth, it is clear that large sums of money for scholarships and fellowships and monetary and other rewards in disproportionate amounts might draw into science too large a percentage of the Nation's high ability, with a result highly detrimental to the Nation and to science. Plans for the discovery and development of scientific talent must be related to the other needs of society for high ability * * *. There is never enough ability at high levels to satisfy all the needs of the Nation; we would not seek to draw into science any more of it than science's proportionate share."

23

The Wartime Deficit

Among the young men and women qualified to take up scientific work, since 1940 there have been few students over 18, except some in medicine and engineering in Army and Navy programs and a few 4-F's, who have followed an integrated scientific course of studies. Neither our allies nor, so far as we know, our enemies have done anything so radical as thus to suspend almost completely their educational activities in scientific pursuits during the war period.

Two great principles have guided us in this country as we have turned our full efforts to war. First, the sound democratic principle that there should be no favored classes or special privilege in a time of peril, that all should be ready to sacrifice equally; second, the tenet that every man should serve in the capacity in which his talents and experience can best be applied for the prosecution of the war effort. In general we have held these principles well in balance.

In my opinion, however, we have drawn too heavily for nonscientific purposes upon the great natural resource which resides in our trained young scientists and engineers. For the general good of the country too many such men have gone into uniform, and their talents have not always been fully utilized. With the exception of those men engaged in war research, all physically fit students at graduate level have been taken into the armed forces. Those ready for college training in the sciences have not been permitted to enter upon that training.

There is thus an accumulating deficit of trained research personnel which will continue for many years. The deficit of science and technology students who, but for the war, would have received bachelor's degrees is about 150,000. The deficit of those holding advanced degrees—that is, young scholars trained to the point where they are capable of carrying on original work—has been estimated as amounting to about 17,000 by 1955 in chemistry, engineering, geology, mathematics, physics, psychology, and the biological sciences.

With mounting demands for scientists both for teaching and for research, we will enter the postwar period with a serious deficit in our trained scientific personnel.

Improve the Quality

Confronted with these deficits, we are compelled to look to the use of our basic human resources and formulate a program which will assure their conservation and effective development. The committee advising me on scientific personnel has stated the following principle which should guide our planning:

"If we were all-knowing and all-wise we might, but we think probably not, write you a plan whereby there might be selected for training, which they otherwise would not get, those who, 20 years hence, would be scientific leaders, and we might not bother about any lesser manifestations of scientific ability. But in the present state of knowledge a plan cannot be made which will select, and assist, only those young men and women who will give the top future leadership to science. To get top leadership there must be a relatively large base of high ability selected for development and then successive skimmings of the cream of ability at successive times and at higher levels. No one can select from the bottom those who will be the leaders at the top because unmeasured and unknown factors enter into scientific, or any, leadership. There are brains and character, strength and health, happiness and spiritual vitality, interest and motivation, and no one knows what else, that must needs enter into this supra-mathematical calculus.

"We think we probably would not, even if we were all-wise and all-knowing, write you a plan whereby you would be assured of scientific leadership at one stroke. We think as we think because we are not interested in setting up an elect. We think it much the best plan, in this constitutional Republic, that opportunity be held out to all kinds and conditions of men whereby they can better themselves. This is the American way; this is the way the United States has become what it is. We think it very important that circumstances be such that there be no ceilings, other than ability itself, to intellectual ambition. We think it very important that every boy and girl shall know that, if he shows that he has what it takes, the sky is the limit. Even if it be shown subsequently that he has not what it takes to go to the top, he will go farther than he would otherwise go if there had been a ceiling beyond which he always knew he could not aspire.

"By proceeding from point to point and taking stock on the way, by giving further opportunity to those who show themselves worthy of further opportunity, by giving the most opportunity to those who show themselves continually developing—this is the way we propose. This is the American way: a man works for what he gets."

Remove the Barriers

Higher education in this country is largely for those who have the means. If those who have the means coincided entirely with those persons who have the talent we should not be squandering a part of our higher education on those undeserving of it, nor neglecting great talent among those who fail to attend college for economic reasons. There are talented individuals in every segment of the population, but with few exceptions those without the means of buying higher education go without it. Here is a tremendous waste of the greatest resource of a nation—the intelligence of its citizens.

If ability, and not the circumstance of family fortune, is made to determine who shall receive higher education in science, then we shall be assured of constantly improving quality at every level of scientific activity.

The Generation in Uniform Must Not Be Lost

We have a serious deficit in scientific personnel partly because the men who would have studied science in the colleges and universities have been serving in the Armed Forces. Many had begun their studies before they went to war. Others with capacity for scientific education went to war after finishing high school. The most immediate prospect of making up some of the deficit in scientific personnel is by salvaging scientific talent from the generation in uniform. For even if we should start now to train the current crop of high school graduates, it would be 1951 before they would complete graduate studies and be prepared for effective scientific research. This fact underlines the necessity of salvaging potential scientists in uniform.

The Armed Services should comb their records for men who, prior to or during the war, have given evidence of talent for science, and make prompt arrangements, consistent with current discharge plans, for ordering those who remain in uniform as soon as militarily possible to duty at institutions

*here and overseas where they can continue their scientific education. More-
over, they should see that those who study overseas have the benefit of the
latest scientific developments.*

A Program

The country may be proud of the fact that 95 percent of boys and girls
of fifth grade age are enrolled in school, but the drop in enrollment after
the fifth grade is less satisfying. For every 1,000 students in the fifth grade,
600 are lost to education before the end of high school, and all but 72 have
ceased formal education before completion of college. While we are concerned primarily with methods of selecting and educating high school graduates at the college and higher levels, we cannot be complacent about the
loss of potential talent which is inherent in the present situation.

Students drop out of school, college, and graduate school, or do not get
that far, for a variety of reasons: they cannot afford to go on; schools and
colleges providing courses equal to their capacity are not available locally;
business and industry recruit many of the most promising before they have
finished the training of which they are capable. These reasons apply with
particular force to science: the road is long and expensive; it extends at least
6 years beyond high school; the percentage of science students who can obtain
first-rate training in institutions near home is small.

Improvement in the teaching of science is imperative, for students of latent
scientific ability are particularly vulnerable to high school teaching which
fails to awaken interest or to provide adequate instruction. To enlarge the
group of specially qualified men and women it is necessary to increase the
number who go to college. This involves improved high school instruction,
provision for helping individual talented students to finish high school
(primarily the responsibility of the local communities), and opportunities
for more capable, promising high school students to go to college. Anything
short of this means serious waste of higher education and neglect of human
resources.

*To encourage and enable a larger number of young men and women of
ability to take up science as a career, and in order gradually to reduce the
deficit of trained scientific personnel, it is recommended that provision be
made for a reasonable number of (a) undergraduate scholarships and graduate fellowships and (b) fellowships for advanced training and fundamental
research. The details should be worked out with reference to the interests
of the several States and of the universities and colleges; and care should
be taken not to impair the freedom of the institutions and individuals
concerned.*

The program proposed by the Moe Committee in Appendix 4 would provide 24,000 undergraduate scholarships and 900 graduate fellowships and
would cost about $30,000,000 annually when in full operation. Each year
under this program 6,000 undergraduate scholarships would be made available to high school graduates, and 300 graduate fellowships would be offered
to college graduates. Approximately the scale of allowances provided for

26

under the educational program for returning veterans has been used in estimating the cost of this program.

The plan is, further, that all those who receive such scholarships or fellowships in science should be enrolled in a National Science Reserve and be liable to call into the service of the Government, in connection with scientific or technical work in time of war or other national emergency declared by Congress or proclaimed by the President. Thus, in addition to the general benefits to the Nation by reason of the addition to its trained ranks of such a corps of scientific workers, there would be a definite benefit to the Nation in having these scientific workers on call in national emergencies. The Government would be well advised to invest the money involved in this plan even if the benefits to the Nation were thought of solely—which they are not—in terms of national preparedness.

Part Five

A PROBLEM OF SCIENTIFIC
RECONVERSION

Effects of Mobilization of Science for War

We have been living on our fat. For more than 5 years many of our scientists have been fighting the war in the laboratories, in the factories and shops, and at the front. We have been directing the energies of our scientists to the development of weapons and materials and methods on a large number of relatively narrow projects initiated and controlled by the Office of Scientific Research and Development and other Government agencies. Like troops, the scientists have been mobilized and thrown into action to serve their country in time of emergency. But they have been diverted to a greater extent than is generally appreciated from the search for answers to the fundamental problems—from the search on which human welfare and progress depends. This is not a complaint—it is a fact. The mobilization of science behind the lines is aiding the fighting men at the front to win the war and to shorten it; and it has resulted incidentally in the accumulation of a vast amount of experience and knowledge of the application of science to particular problems, much of which can be put to use when the war is over. Fortunately, this country had the scientists—and the time—to make this contribution and thus to advance the date of victory.

Security Restrictions Should be Lifted Promptly

Much of the information and experience acquired during the war is confined to the agencies that gathered it. Except to the extent that military security dictates otherwise, such knowledge should be spread upon the record for the benefit of the general public.

Thanks to the wise provision of the Secretary of War and the Secretary of the Navy, most of the results of wartime medical research have been published. Several hundred articles have appeared in the professional journals; many are in process of publication. The material still subject to security classification should be released as soon as possible.

It is my view that most of the remainder of the classified scientific material should be released as soon as there is ground for belief that the enemy will not be able to turn it against us in this war. Most of the information needed by industry and in education can be released without disclosing its embodiments in actual military material and devices. Basically there is no reason to believe that scientists of other countries will not in time rediscover everything we now know which is held in secrecy. A broad dissemination of scientific information upon which further advances can readily be made furnishes a sounder foundation for our national security than a policy of restriction which would impede our own progress although imposed in the hope that possible enemies would not catch up with us.

During the war it has been necessary for selected groups of scientists to work on specialized problems, with relatively little information as to what other groups were doing and had done. Working against time, the Office of Scientific Research and Development has been obliged to enforce this practice during the war, although it was realized by all concerned that it was an emergency measure which prevented the continuous cross-fertilization so essential to fruitful scientific effort.

Our ability to overcome possible future enemies depends upon scientific advances which will proceed more rapidly with diffusion of knowledge than under a policy of continued restriction of knowledge now in our possession.

Need for Coordination

In planning the release of scientific data and experience collected in connection with the war, we must not overlook the fact that research has gone forward under many auspices—the Army, the Navy, the Office of Scientific Research and Development, the National Advisory Committee for Aeronautics, other departments and agencies of the Government, educational institutions, and many industrial organizations. There have been numerous cases of independent discovery of the same truth in different places. To permit the release of information by one agency and to continue to restrict it elsewhere would be unfair in its effect and would tend to impair the morale and efficiency of scientists who have submerged individual interests in the controls and restrictions of war.

A part of the information now classified which should be released is possessed jointly by our allies and ourselves. Plans for release of such information should be coordinated with our allies to minimize danger of international friction which would result from sporadic uncontrolled release.

A Board to Control Release

The agency responsible for recommending the release of information from military classification should be an Army, Navy, civilian body, well grounded in science and technology. It should be competent to advise the Secretary of War and the Secretary of the Navy. It should, moreover, have sufficient recognition to secure prompt and practical decisions.

To satisfy these considerations I recommend the establishment of a Board, made up equally of scientists and military men, whose function would be to pass upon the declassification and to control the release for publication of scientific information which is now classified.

29

Publication Should Be Encouraged

The release of information from security regulations is but one phase of the problem. The other is to provide for preparation of the material and its publication in a form and at a price which will facilitate dissemination and use. In the case of the Office of Scientific Research and Development, arrangements have been made for the preparation of manuscripts, while the staffs under our control are still assembled and in possession of the records, as soon as the pressure for production of results for this war has begun to relax.

We should get this scientific material to scientists everywhere with great promptness, and at as low a price as is consistent with suitable format. We should also get it to the men studying overseas so that they will know what has happened in their absence.

It is recommended that measures which will encourage and facilitate the preparation and publication of reports be adopted forthwith by all agencies, governmental and private, possessing scientific information released from security control.

Part Six

THE MEANS TO THE END

New Responsibilities for Government

One lesson is clear from the reports of the several committees attached as appendices. The Federal Government should accept new responsibilities for promoting the creation of new scientific knowledge and the development of scientific talent in our youth.

The extent and nature of these new responsibilities are set forth in detail in the reports of the committees whose recommendations in this regard are fully endorsed.

In discharging these responsibilities Federal funds should be made available. We have given much thought to the question of how plans for the use of Federal funds may be arranged so that such funds will not drive out of the picture funds from local governments, foundations, and private donors. We believe that our proposals will minimize that effect, but we do not think that it can be completely avoided. We submit, however, that the Nation's need for more and better scientific research is such that the risk must be accepted.

It is also clear that the effective discharge of these responsibilities will require the full attention of some over-all agency devoted to that purpose. There should be a focal point within the Government for a concerted program of assisting scientific research conducted outside of Government. Such an agency should furnish the funds needed to support basic research in the colleges and universities, should coordinate where possible research programs on matters of utmost importance to the national welfare, should formulate a national policy for the Government toward science, should sponsor the interchange of scientific information among scientists and laboratories both in this country and abroad, and should ensure that the incentives to research in industry and the universities are maintained. All of the committees advising on these matters agree on the necessity for such an agency.

The Mechanism

There are within Government departments many groups whose interests are primarily those of scientific research. Notable examples are found within the Departments of Agriculture, Commerce, Interior, and the Federal Security Agency. These groups are concerned with science as collateral and peripheral to the major problems of those Departments. These groups should remain where they are, and continue to perform their present functions, including the support of agricultural research by grants to the land grant colleges and experimental stations, since their largest contribution lies in applying fundamental knowledge to the special problems of the Departments within which they are established.

By the same token these groups cannot be made the repository of the new and large responsibilities in science which belong to the Government and which the Government should accept. The recommendations in this report which relate to research within the Government, to the release of scientific information, to clarification of the tax laws, and to the recovery and development of our scientific talent now in uniform can be implemented by action within the existing structure of the Government. But nowhere in the governmental structure receiving its funds from Congress is there an agency adapted to supplementing the support of basic research in the universities, both in medicine and the natural sciences; adapted to supporting research on new weapons for both Services; or adapted to administering a program of science scholarships and fellowships.

A new agency should be established, therefore, by the Congress for the purpose. Such an agency, moreover, should be an independent agency devoted to the support of scientific research and advanced scientific education alone. Industry learned many years ago that basic research cannot often be fruitfully conducted as an adjunct to or a subdivision of an operating agency or department. Operating agencies have immediate operating goals and are under constant pressure to produce in a tangible way, for that is the test of their value. None of these conditions is favorable to basic research. Research is the exploration of the unknown and is necessarily speculative. It is inhibited by conventional approaches, traditions, and standards. It cannot be satisfactorily conducted in an atmosphere where it is gauged and tested by operating or production standards. Basic scientific research should not, therefore, be placed under an operating agency whose paramount concern is anything other than research. Research will always suffer when put in competition with operations. The decision that there should be a new and independent agency was reached by each of the committees advising in these matters.

I am convinced that these new functions should be centered in one agency. Science is fundamentally a unitary thing. The number of independent agencies should be kept to a minimum. Much medical progress, for example, will come from fundamental advances in chemistry. Separation of the sciences in tight compartments, as would occur if more than one agency were involved, would retard and not advance scientific knowledge as a whole.

Five Fundamentals

There are certain basic principles which must underlie the program of Government support for scientific research and education if such support is to be effective and if it is to avoid impairing the very things we seek to foster. These principles are as follows:

32

(1) Whatever the extent of support may be, there must be stability of funds over a period of years so that long-range programs may be undertaken.

(2) The agency to administer such funds should be composed of citizens selected only on the basis of their interest in and capacity to promote the work of the agency. They should be persons of broad interest in and understanding of the peculiarities of scientific research and education.

(3) The agency should promote research through contracts or grants to organizations outside the Federal Government. It should not operate any laboratories of its own.

(4) Support of basic research in the public and private colleges, universities, and research institutes must leave the internal control of policy, personnel, and the method and scope of the research to the institutions themselves. This is of the utmost importance.

(5) While assuring complete independence and freedom for the nature, scope, and methodology of research carried on in the institutions receiving public funds, and while retaining discretion in the allocation of funds among such institutions, the Foundation proposed herein must be responsible to the President and the Congress. Only through such responsibility can we maintain the proper relationship between science and other aspects of a democratic system. The usual controls of audits, reports, budgeting, and the like, should, of course, apply to the administrative and fiscal operations of the Foundation, subject, however, to such adjustments in procedure as are necessary to meet the special requirements of research.

Basic research is a long-term process—it ceases to be basic if immediate results are expected on short-term support. Methods should therefore be found which will permit the agency to make commitments of funds from current appropriations for programs of five years duration or longer. Continuity and stability of the program and its support may be expected (a) from the growing realization by the Congress of the benefits to the public from scientific research, and (b) from the conviction which will grow among those who conduct research under the auspices of the agency that good quality work will be followed by continuing support.

Military Research

As stated earlier in this report, military preparedness requires a permanent, independent, civilian-controlled organization, having close liaison with the Army and Navy, but with funds direct from Congress and the clear power to initiate military research which will supplement and strengthen that carried on directly under the control of the Army and Navy. As a temporary measure the National Academy of Sciences has established the Research Board for National Security at the request of the Secretary of War and the Secretary of the Navy. This is highly desirable in order that there may be no interruption in the relations between scientists and military men after the emergency wartime Office of Scientific Research and Development goes out of existence. The Congress is now considering legislation to provide funds for this Board by direct appropriation.

I believe that, as a permanent measure, it would be appropriate to add to the agency needed to perform the other functions recommended in this report the responsibilities for civilian-initiated and civilian-controlled military research. The function of such a civilian group would be primarily to conduct long-range scientific research on military problems—leaving to the Services research on the improvement of existing weapons.

Some research on military problems should be conducted, in time of peace as well as in war, by civilians independently of the military establishment. It is the primary responsibility of the Army and Navy to train the men, make available the weapons, and employ the strategy that will bring victory in combat. The Armed Services cannot be expected to be experts in all of the complicated fields which make it possible for a great nation to fight successfully in total war. There are certain kinds of research—such as research on the improvement of existing weapons—which can best be done within the military establishment. However, the job of long-range research involving application of the newest scientific discoveries to military needs should be the responsibility of those civilian scientists in the universities and in industry who are best trained to discharge it thoroughly and successfully. It is essential that both kinds of research go forward and that there be the closest liaison between the two groups.

Placing the civilian military research function in the proposed agency would bring it into close relationship with a broad program of basic research in both the natural sciences and medicine. A balance between military and other research could thus readily be maintained.

The establishment of the new agency, including a civilian military research group, should not be delayed by the existence of the Research Board for National Security, which is a temporary measure. Nor should the creation of the new agency be delayed by uncertainties in regard to the postwar organization of our military departments themselves. Clearly, the new agency, including a civilian military research group within it, can remain sufficiently flexible to adapt its operations to whatever may be the final organization of the military departments.

National Research Foundation

It is my judgment that the national interest in scientific research and scientific education can best be promoted by the creation of a National Research Foundation.

I. Purposes

The National Research Foundation should develop and promote a national policy for scientific research and scientific education, should support basic research in nonprofit organizations, should develop scientific talent in American youth by means of scholarships and fellowships, and should by contract and otherwise support long-range research on military matters.

II. Members

1. Responsibility to the people, through the President and Congress, should be placed in the hands of, say nine Members, who should be persons not

otherwise connected with the Government and not representative of any special interest, who should be known as National Research Foundation Members, selected by the President on the basis of their interest in and capacity to promote the purposes of the Foundation.

2. The terms of the Members should be, say, 4 years, and no Member should be eligible for immediate reappointment provided he has served a full 4-year term. It should be arranged that the Members first appointed serve terms of such length that at least two Members are appointed each succeeding year.

3. The Members should serve without compensation but should be entitled to their expenses incurred in the performance of their duties.

4. The Members should elect their own chairman annually.

5. The chief executive officer of the Foundation should be a director appointed by the Members. Subject to the direction and supervision of the Foundation Members (acting as a board), the director should discharge all the fiscal, legal, and administrative functions of the Foundation. The director should receive a salary that is fully adequate to attract an outstanding man to the post.

6. There should be an administrative office responsible to the director to handle in one place the fiscal, legal, personnel, and other similar administrative functions necessary to the accomplishment of the purposes of the Foundation.

7. With the exception of the director, the division members, and one executive officer appointed by the director to administer the affairs of each division, all employees of the Foundation should be appointed under Civil Service regulations.

III. *Organization*

1. In order to accomplish the purposes of the Foundation, the Members should establish several professional Divisions to be responsible to the Members. At the outset these Divisions should be:

   a. *Division of Medical Research.*—The function of this Division should be to support medical research.

   b. *Division of Natural Sciences.*—The function of this Division should be to support research in the physical and natural sciences.

   c. *Division of National Defense.*—It should be the function of this Division to support long-range scientific research on military matters.

   d. *Division of Scientific Personnel and Education.*—It should be the function of this Division to support and to supervise the grant of scholarships and fellowships in science.

   e. *Division of Publications and Scientific Collaboration.*—This Division should be charged with encouraging the publication of scientific knowledge and promoting international exchange of scientific information.

2. Each Division of the Foundation should be made up of at least five members, appointed by the Members of the Foundation. In making such appointments the Members should request and consider recommendations from the National Academy of Sciences which should be asked to establish a new National Research Foundation nominating committee in order to

bring together the recommendations of scientists in all organizations. The chairman of each Division should be appointed by the Members of the Foundation.

3. The Division Members should be appointed for such terms as the Members of the Foundation may determine, and may be reappointed at the discretion of the Members. They should receive their expenses and compensation for their services at a per diem rate of, say, $50 while engaged on business of the Foundation, but no Division Member should receive more than, say, $10,000 compensation per year.

4. Membership of the Division of National Defense should include, in addition to, say, five civilian members, one representative designated by the Secretary of War, and one representative of the Secretary of the Navy, who should serve without additional compensation for this duty.

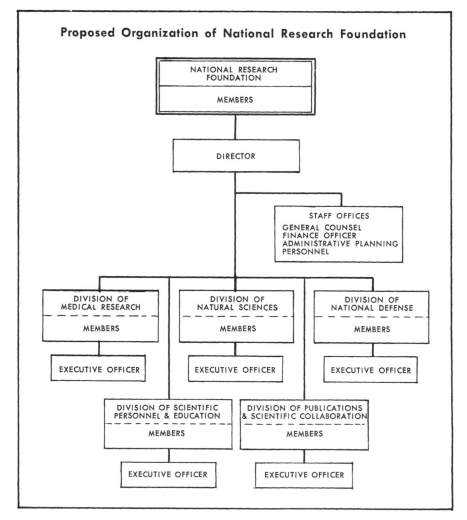

Proposed Organization of National Research Foundation

IV. *Functions*

1. *The Members of the Foundation should have the following functions, powers, and duties:*

a. To formulate over-all policies of the Foundation.

b. To establish and maintain such offices within the United States, its territories and possessions, as they may deem necessary.

c. To meet and function at any place within the United States, its territories and possessions.

d. To obtain and utilize the services of other Government agencies to the extent that such agencies are prepared to render such services.

e. To adopt, promulgate, amend, and rescind rules and regulations to carry out the provisions of the legislation and the policies and practices of the Foundation.

f. To review and balance the financial requirements of the several Divisions and to propose to the President the annual estimate for the funds required by each Division. Appropriations should be earmarked for the purposes of specific Divisions, but the Foundation should be left discretion with respect to the expenditure of each Division's funds.

g. To make contracts or grants for the conduct of research by negotiation without advertising for bids.

And with the advice of the National Research Foundation Divisions concerned—

h. To create such advisory and cooperating agencies and councils, State, regional, or national, as in their judgment will aid in effectuating the purposes of the legislation, and to pay the expenses thereof.

i. To enter into contracts with or make grants to educational and non-profit research institutions for support of scientific research.

j. To initiate and finance in appropriate agencies, institutions, or organizations, research on problems related to the national defense.

k. To initiate and finance in appropriate organizations research projects for which existing facilities are unavailable or inadequate.

l. To establish scholarships and fellowships in the natural sciences including biology and medicine.

m. To promote the dissemination of scientific and technical information and to further its international exchange.

n. To support international cooperation in science by providing financial aid for international meetings, associations of scientific societies, and scientific research programs organized on an international basis.

o. To devise and promote the use of methods of improving the transition between research and its practical application in industry.

2. *The Divisions should be responsible to the Members of the Foundation for—*

a. Formulation of programs and policy within the scope of the particular Divisions.

b. Recommendation regarding the allocation of research programs among research organizations.

c. Recommendation of appropriate arrangements between the Foundation and the organizations selected to carry on the program.

d. Recommendation of arrangements with State and local authorities in regard to cooperation in a program of science scholarships and fellowships.

e. Periodic review of the quality of research being conducted under the auspices of the particular Division and revision of the program of support of research.

f. Presentation of budgets of financial needs for the work of the Division.

g. Maintaining liaison with other scientific research agencies, both governmental and private, concerned with the work of the Division.

V. *Patent Policy*

The success of the National Research Foundation in promoting scientific research in this country will depend to a very large degree upon the cooperation of organizations outside the Government. In making contracts with or grants to such organizations the Foundation should protect the public interest adequately and at the same time leave the cooperating organizations with adequate freedom and incentive to conduct scientific research. The public interest will normally be adequately protected if the Government receives a royalty-free license for governmental purposes under any patents resulting from work financed by the Foundation. There should be no obligation on the research institution to patent discoveries made as a result of support from the Foundation. There should certainly *not* be any absolute requirement that all rights in such discoveries be assigned to the Government, but it should be left to the discretion of the Director and the interested Division whether in special cases the public interest requires such an assignment. Legislation on this point should leave to the Members of the Foundation discretion as to its patent policy in order that patent arrangements may be adjusted as circumstances and the public interest require.

VI. *Special Authority*

In order to insure that men of great competence and experience may be designated as Members of the Foundation and as Members of the several professional Divisions, the legislation creating the Foundation should contain specific authorization so that the Members of the Foundation and the Members of the Divisions may also engage in private and gainful employment, notwithstanding the provisions of any other laws: provided, however, that no compensation for such employment is received in any form from any profit-making institution which receives funds under contract, or otherwise, from the Division or Divisions of the Foundation with which the individual is concerned. In normal times, in view of the restrictive statutory prohibitions against dual interests on the part of Government officials, it would be virtually impossible to persuade persons having private employment of any kind to serve the Government in an official capacity. In order, however, to secure the part-time services of the most competent men as Members of the Foundation and the Divisions, these stringent prohibitions should be relaxed to the extent indicated.

Since research is unlike the procurement of standardized items, which are susceptible to competitive bidding on fixed specifications, the legislation creating the National Research Foundation should free the Foundation from the obligation to place its contracts for research through advertising for bids. This is particularly so since the measure of a successful research contract lies not in the dollar cost but in the qualitative and quantitative contribution which is made to our knowledge. The extent of this contribution in turn depends on the creative spirit and talent which can be brought to bear within a research laboratory. The National Research Foundation must, therefore, be free to place its research contracts or grants not only with those institutions which have a demonstrated research capacity but also with other institutions whose latent talent or creative atmosphere affords promise of research success.

As in the case of the research sponsored during the war by the Office of Scientific Research and Development, the research sponsored by the National Research Foundation should be conducted, in general, on an actual cost basis without profit to the institution receiving the research contract or grant.

There is one other matter which requires special mention. Since research does not fall within the category of normal commercial or procurement operations which are easily covered by the usual contractual relations, it is essential that certain statutory and regulatory fiscal requirements be waived in the case of research contractors. For example, the National Research Foundation should be authorized by legislation to make, modify, or amend contracts of all kinds with or without legal consideration, and without performance bonds. Similarly, advance payments should be allowed in the discretion of the Director of the Foundation when required. Finally, the normal vouchering requirements of the General Accounting Office with respect to detailed itemization or substantiation of vouchers submitted under cost contracts should be relaxed for research contractors. Adherence to the usual procedures in the case of research contracts will impair the efficiency of research operations and will needlessly increase the cost of the work to the Government. Without the broad authority along these lines which was contained in the First War Powers Act and its implementing Executive Orders, together with the special relaxation of vouchering requirements granted by the General Accounting Office, the Office of Scientific Research and Development would have been gravely handicapped in carrying on research on military matters during this war. Colleges and universities in which research will be conducted principally under contract with the Foundation are, unlike commercial institutions, not equipped to handle the detailed vouchering procedures and auditing technicalities which are required of the usual Government contractors.

VII. *Budget*

Studies by the several committees provide a partial basis for making an estimate of the order of magnitude of the funds required to implement the proposed program. Clearly the program should grow in a healthy manner from modest beginnings. The following very rough estimates are given for the first year of operation after the Foundation is organized and operating, and for the fifth year of operation when it is expected that the operations would have reached a fairly stable level:

Activity	Millions of dollars	
	First year	Fifth year
Division of Medical Research	$5.0	$20.0
Division of Natural Sciences	10.0	50.0
Division of National Defense	10.0	20.0
Division of Scientific Personnel and Education	7.0	29.0
Division of Publications and Scientific Collaboration	.5	1.0
Administration	1.0	2.5
	33.5	122.5

Action by Congress

The National Research Foundation herein proposed meets the urgent need of the days ahead. The form of the organization suggested is the result of considerable deliberation. The form is important. The very successful pattern of organization of the National Advisory Committee for Aeronautics, which has promoted basic research on problems of flight during the past thirty years, has been carefully considered in proposing the method of appointment of Members of the Foundation and in defining their responsibilities. Moreover, whatever program is established it is vitally important that it satisfy the Five Fundamentals.

The Foundation here proposed has been described only in outline. The excellent reports of the committees which studied these matters are attached as appendices. They will be of aid in furnishing detailed suggestions.

Legislation is necessary. It should be drafted with great care. Early action is imperative, however, if this Nation is to meet the challenge of science and fully utilize the potentialities of science. On the wisdom with which we bring science to bear against the problems of the coming years depends in large measure our future as a Nation.

APPENDICES

IN MEMORIAM

The following members of the advisory committees have died since the publication of *Science, the Endless Frontier* in 1945:

Isaiah Bowman

Walter C. Coffey

Karl T. Compton

R. E. Doherty

Clarence A. Dykstra

Farnham P. Griffiths

W. S. Hunter

W. Rupert Maclaurin

Charles E. MacQuigg

Cleveland Norcross

J. Hugh O'Donnell

W. W. Palmer

J. T. Tate

Kenneth B. Turner

Committees Consulted

Question

"With particular reference to the war of science against disease, what can be done now to organize a program for continuing in the future the work which has been done in medicine and related sciences?"

Committee

Dr. W. W. Palmer, chairman; Bard professor of medicine, Columbia University; director of medical service of Presbyterian Hospital, New York City.

Dr. Homer W. Smith, secretary; director, physiology laboratory, School of Medicine, New York University.

Dr. Kenneth B. Turner, assistant secretary; assistant professor of medicine, Columbia University.

Dr. W. B. Castle, professor of medicine, Harvard University; associate director, Thorndike Memorial Laboratory, Boston City Hospital.

Dr. Edward A. Doisy, director, department of physiology and biochemistry, St. Louis University School of Medicine (recipient of Nobel Award).

Dr. Ernest Goodpasture, professor of pathology, School of Medicine, Vanderbilt University.

Dr. Alton Ochsner, professor of surgery and head of the department of surgery at Tulane University School of Medicine.

Dr. Linus Pauling, head of the division of chemistry and chemical engineering and director of the chemical laboratories at the California Institute of Technology.

Dr. James J. Waring, professor of medicine, University of Colorado School of Medicine.

Question

"What can the Government do now and in the future to aid research activities by public and private organizations? The proper roles of public and of private research, and their interrelation, should be carefully considered."

Committee

Dr. Isaiah Bowman, chairman; president of Johns Hopkins University.

Dr. J. T. Tate, vice chairman; research professor of physics, University of Minnesota.

Dr. W. Rupert Maclaurin, secretary; professor of economics, Massachusetts Institute of Technology.

Dr. Oliver E. Buckley, president of the Bell Telephone Laboratories.

Dr. Walter C. Coffey, president of the University of Minnesota.

Mr. Oscar S. Cox, deputy administrator of the Foreign Economic Administration.

Col. Bradley Dewey, president of Dewey & Almy Chemical Co.

Dr. Clarence A. Dykstra, provost of the University of California at Los Angeles.

Dr. C. P. Haskins, director of Haskins Laboratories.

Dr. Edwin H. Land, president and director of research, Polaroid Corporation.

Dr. Charles E. MacQuigg, dean of the College of Engineering, Ohio State University.

Dr. Harold G. Moulton, president of the Brookings Institution.

Rev. J. Hugh O'Donnell, president of the University of Notre Dame.

Dr. I. I. Rabi, professor of physics, Columbia University (recipient of Nobel Award).

Dr. Warren Weaver, director for natural sciences, Rockefeller Foundation.

Dr. Robert E. Wilson, chairman of the board, Standard Oil Co. of Indiana.

Dr. William E. Wrather, director, U. S. Geological Survey.

Question

"Can an effective program be proposed for discovering and developing scientific talent in American youth so that the continuing future of scientific research in this country may be assured on a level comparable to what has been done during the war?"

Committee

Dr. Henry Allen Moe, chairman; secretary-general of the John Simon Guggenheim Memorial Foundation.

Mr. Lawrence K. Frank, secretary.

Mr. Henry Chauncey, assistant secretary.

Dr. Henry A. Barton, director of the American Institute of Physics.

Dr. C. Lalor Burdick, special assistant to the president, E. I. du Pont de Nemours & Co.

Continued, next page

Dr. J. B. Conant, president of Harvard University; chairman of the National Defense Research Committee.

Dr. Watson Davis, editor and director of Science Service.

Dr. R. E. Doherty, president of the Carnegie Institute of Technology.

Dr. Paul E. Elicker, executive secretary, National Association of Secondary School Principals.

Mr. Farnham P. Griffiths, lawyer, San Francisco.

Dr. W. S. Hunter, professor of psychology at Brown University.

Dr. T. R. McConnell, dean of the College of Science, Literature, and Arts at the University of Minnesota.

Mr. Walter S. Rogers, director of the Institute of Current World Affairs.

Dr. Harlow Shapley, director of the Harvard College Observatory.

Dr. Hugh S. Taylor, dean of the Graduate School, Princeton University.

Dr. E. B. Wilson, professor of vital statistics, Harvard University School of Public Health.

Question

"What can be done, consistent with military security, and with the prior approval of the military authorities, to make known to the world as soon as possible the contributions which have been made during our war effort to scientific knowledge?"

Committee

Dr. Irvin Stewart, chairman; executive secretary of the Office of Scientific Research and Development; director of the Committee on Scientific Aids to Learning of the National Research Council.

Mr. Cleveland Norcross, secretary; executive assistant to the executive secretary of the Office of Scientific Research and Development.

Dr. J. P. Baxter III, president of Williams College; historian of the Office of Scientific Research and Development.

Dr. Karl T. Compton, president of the Massachusetts Institute of Technology; chairman of the Research Board for National Security; member of the National Defense Research Committee.

Dr. J. B. Conant, president of Harvard University; chairman of the National Defense Research Committee.

Dr. A. N. Richards, vice president of the University of Pennsylvania in charge of Medical Affairs; chairman of the Committee on Medical Research of the Office of Scientific Research and Development.

Dr. M. A. Tuve, director, applied physics laboratory, Johns Hopkins University; staff member of the department of Terrestrial Magnetism of the Carnegie Institution of Washington.

Mr. Carroll L. Wilson, executive assistant to the Director of the Office of Scientific Research and Development.

Appendix 2

Report of the Medical
Advisory Committee

Table of Contents

LETTER OF TRANSMITTAL

Dr. Vannevar Bush, *Director,*
Office of Scientific Research and Development,
1530 P Street NW., Washington 25, D. C.

My Dear Dr. Bush:

It is my privilege to submit herewith the report of the Medical Advisory Committee appointed by you in January of this year to answer the second question in President Roosevelt's letter of November 17, 1944, which was worded:

With particular reference to the war of science against disease, what can be done now to organize a program for continuing in the future the work which has been done in medicine and related sciences? The fact that the annual deaths in this country from one or two diseases alone are far in excess of the total number of lives lost by us in battle during this war should make us conscious of the duty we owe future generations.

In preparing this report, the Committee has consulted some 350 representatives from 73 of the 77 medical schools of the United States, from the Services, from various research institutions, from the pharmaceutical industry, and from philanthropic foundations; it has conferred in joint meeting with the Committee on Medical Research; and it has received written comment and advice from many leaders in medicine and allied sciences throughout the country.

The report, which is preceded by a summary, is in three parts: (1) Considerations on which the recommendations of the Committee are based, (2) fundamental principles governing the use of Federal funds for medical research, (3) recommendations outlining the establishment of a National Foundation for Medical Research as an independent Federal agency.

The Committee recognizes a great and urgent need for the expansion and renovation of medical school laboratories. However, our study has taken no account of this requirement, pertinent as it is to medical research, since a building program was considered outside the scope of our assignment.

This report has the unanimous approval of my Committee and I submit it with the conviction that it has, almost without exception, the endorsement of the many individuals to whom the Committee is so deeply indebted for freely given and valuable advice.

Respectfully yours,

Walter W. Palmer, *Chairman,*
Medical Advisory Committee.

April 25, 1945.

47

MEMBERS OF THE COMMITTEE

Dr. Walter W. Palmer, chairman, Bard professor of medicine, Columbia University; director of medical service of the Presbyterian Hospital, New York City.

Dr. Homer W. Smith, secretary, director, physiology laboratory, School of Medicine, New York University.

Dr. Kenneth B. Turner, assistant secretary, assistant professor of medicine, Columbia University.

Dr. William B. Castle, professor of medicine, Harvard University; associate director, Thorndike Memorial Laboratory, Boston City Hospital.

Dr. Edward A. Doisy, director, department of physiology and biochemistry, St. Louis University School of Medicine (recipient of Nobel Award).

Dr. Ernest Goodpasture, professor of pathology, School of Medicine, Vanderbilt University.

Dr. Alton Ochsner, professor of surgery and head of the department of surgery, Tulane University School of Medicine.

Dr. Linus Pauling, head of the division of chemistry and chemical engineering, director of the chemical laboratories at the California Institute of Technology.

Dr. James J. Waring, professor of medicine, University of Colorado School of Medicine.

SUMMARY

Impressed by the contributions medicine has made in the present world struggle, President Roosevelt asked what could be done by the Government in the future to aid "the war of science against disease."

Recognition of the brilliant record of medicine in World War II has brought comfort to thousands of families with members in the armed forces. Compared to World War I the death rate for all diseases in the Army, including overseas forces, has fallen from 14.1 to 0.6 per 1,000 strength. Penicillin and the sulfonamides, the insecticide DDT, better vaccines, and improved hygienic measures have all but conquered yellow fever, dysentery, typhus, tetanus, pneumonia, meningitis. Malaria has been controlled. Disability from venereal disease has been radically reduced by new methods of treatment. Dramatic progress in surgery has been aided by the increased availability of blood and plasma for transfusions.

Much of the credit for these advances is properly assignable to the Committee on Medical Research of the Office of Scientific Research and Development. In 3 years this organization has developed penicillin and DDT; supported blood fractionation studies resulting in serum albumin as a blood substitute and immune globulin as a new countermeasure against infections; and standardized the effective treatment of malaria with atabrine now used by the armed forces. Up to July 1944, this program had cost $15,000,000, a modest outlay for the saving in suffering and lives.

These dramatic advances in medicine during the war have been the result of developmental rather than fundamental research, and have come through the application, to problems of wartime importance, of a large backlog of scientific data accumulated through careful research in the years prior to the war.

In the meantime, sorely needed additions to·basic knowledge have been prevented. The war has forced us to set aside fundamental research to a large extent. Our capacity to carry out research in the future has been impaired by the curtailment of medical education, the absorption of physicians into the armed forces, the prohibition against training draft-eligible men in the basic medical sciences, and the diversion into developmental problems of those scientists who were able to remain in their laboratories.

The universities are the chief contributors to pure science, for research thrives best in an atmosphere of academic freedom. It is to the universities that we must turn to train more men for research and to provide the information that will enable us to solve the problems of cancer, degenerative disease and the ageing process, neuropsychiatric disorders, peptic ulcer, asthma, and even the common cold.

University funds that can be used for medical research are decreasing as research costs rise. Income from endowment is steadily shrinking, while endowment itself is no longer being increased by large new gifts. Medical schools must continue to meet relatively fixed expenses of teaching and overhead from smaller budgets, with the result that less money is left for research.

Medical research will continue in the future, regardless of any adverse circumstances. The Government, however, has an opportunity to play an important role in supplementing the depleted research budgets of medical schools. Federal aid will increase the volume of medical research; it will strengthen the promise of important discovery and speed its fulfillment; it will encourage and develop the financially weaker schools now at a serious disadvantage; and it will enable the United States to maintain its position of world leadership in medical research in competition with the nations of Europe where State funds have long been available for scientific research. When a government wisely invests the people's money in medical research, the people receive huge dividends in the form of better health and longer lives.

If Federal funds are to be used to aid medical research, they should be provided in three forms:

1. Funds should be made available as unrestricted grants, with no portion earmarked for a specific purpose, to supply technical help and materials; to enable a limited number of young people to obtain research experience during their regular course in medicine; to build up research in institutions where, for financial reasons, it is not now well-developed; and to cover a multitude of research requirements within each institution. The administration of these funds should be decentralized to the fullest possible extent, allowing full play to the wisdom and experience of medical school faculties and administrators. If a central agency were to attempt to underwrite a program of this sort item by item, the costs of administration would be prohibitive, and the organization would be too rigid and ponderous to meet the numerous, diverse, and sometimes rapidly varying needs of the institutions.

2. Funds should be made available to support fellowships in order that young people with aptitude for research may be selected, trained, and given an opportunity to carry on research.

3. Funds should be made available to support special projects of considerable magnitude and importance by grants-in-aid.

The Federal agency should receive its funds by such means as to permit it to favor long term grants, up to 10 years.

Federal aid should be initiated modestly. Funds exceeding the capacity of the Nation's research institutions to utilize them effectively would do harm by encouraging mediocre work and by driving away university and foundation support. The responsible agency must remain free from political influence and resistant to special pressures. Its policies must be determined by scientists who bring sympathetic understanding to the problems of research. The agency must have the necessary flexibility to modify its procedures in the light of experience.

From available information it is estimated that approximately 5 to 7 million dollars annually could be used effectively in the immediate postwar period. As the research program develops a larger sum may be required.

Recommendations

The Committee recommends that Government aid be provided for medical research through the creation of an independent Federal agency to be called the National Foundation for Medical Research. The Foundation would consist of a board of trustees, a technical board, and the necessary administrative organization.

The board of trustees would consist of five eminent scientists appointed by the President with the approval of the Senate for terms of 5 years, and in such a way initially as to secure rotation by the retirement of one member each year. The trustees would serve on a part-time basis, be paid for their work, and be appointed without regard to civil-service laws. Meetings of the trustees would be held monthly with one meeting annually in each of six geographical regions. The trustees would determine the policies of the Foundation and act on all requests for funds.

The technical board would consist of 12 scientists, representing special fields of medical science, appointed by the trustees for terms of 3 years, and in such a way initially as to secure rotation by the retirement of 4 members each year. Technical board members would serve on a part-time basis, be paid for their work, and be appointed without regard to the civil-service laws. Members of the technical board would have the necessary aides and ad hoc committees to assist them in the performance of their duties. The technical board would forward all requests for funds to the trustees with recommendations for appropriate action, follow the progress of work supported by the Foundation, and prepare reports or appraisals requested by the trustees.

The financial and other business affairs of the Foundation would be in charge of a full-time executive secretary responsible to the trustees.

The Foundation would not engage in research but would initiate and coordinate research in existing institutions and maintain liaison with interested domestic and foreign agencies.

Part One

Considerations on which the Recommendations of the Committee are Based

1. *The Record of Medicine in World War II*

We believe that at no time has superior medical and surgical care been available to the public generally than is now received by our armed forces even in the most remote parts of the world. Public knowledge of the excellence of this care has brought comfort to thousands of anxious families and has strengthened the morale of our fighting men.

The magnificent records of the medical departments of the Army and Navy are directly attributable to two factors: (1) The training men received before the war in American medical schools and teaching hospitals was the best in the world, and, when war came, large reserves of superbly trained physicians and surgeons were available for the armed forces. (2) Medical progress had been rapid before the war and was continued at an accelerated rate during the war under the stimulus of the Committee on Medical Research and the Army Epidemiology Board.

The results are spectacular. Between World War I and World War II, the death rate for all diseases in the Army, including overseas forces, has been reduced from 14.1 to 0.6 per 1,000 strength. Dysentery, formerly the scourge of armies, has become a minor problem. Tetanus, typhoid, paratyphoid, cholera, and smallpox have been practically eliminated. As a result of a potent vaccine and improved mosquito control, yellow fever has not appeared in the Army or Navy. The prompt arrest of the Naples epidemic of typhus by means of the insecticide DDT is a dramatic example of preventive medicine.

The use of the sulfa drugs has lowered the death rate from lobar pneumonia in the Army from 24 percent in World War I to less than 1 percent at present. The death rate from meningitis has been reduced to one-tenth of that in World War I.

Penicillin is one of the great triumphs of modern therapeutics. By its use death rates and disability from infections due to the staphylococcus, streptococcus, pneumococcus, and anthrax bacillus have been greatly reduced. It has also proved to be a most effective weapon in limiting infection and in accelerating healing of wounds and burns. As a result of treatment with penicillin the days per man per year lost from active duty in 1944 because of venereal disease were one-third of those for 1940. The temporary disabling complications of gonorrhea have been cut in this period to one twenty-fourth.

Advances in surgery have been scarcely less dramatic. Despite devastating antipersonnel munitions, the fatality rate among the wounded has been as low as in any war in history. Prolonged and difficult operations are performed successfully in field hospitals close to the front. Surgical skill has been aided by the availability of large quantities of plasma and whole blood for the treatment of severely wounded men.

2. The Committee on Medical Research of the Office of Scientific Research and Development

In the summer of 1940, the advice of the Division of Medical Sciences of the National Research Council was sought by the Surgeons General in many fields of medicine and surgery. Ultimately 13 committees and 43 subcommittees were set up in aviation medicine, chemotherapy, convalescence and rehabilitation, drugs and medical supplies, industrial medicine, medicine including malarial studies, infectious diseases, nutrition, tropical disease, tuberculosis, venereal diseases, etc., neuropsychiatry, pathology, sanitary engineering, shock and transfusion, surgery, and the treatment of gas casualties.

In June 1941, the Committee on Medical Research was organized under the Office of Scientific Research and Development, to "initiate and support scientific research on medical problems affecting the national defense." The existing committees of the National Research Council acted in an advisory capacity to the new organization.

As of December 1, 1944, 496 research contracts had been executed by the Committee on Medical Research with 120 different institutions. Over 95 percent of these contracts were with universities or teaching hospitals. The personnel represented in this work numbered about 2,670, of whom 553 were physicians. These investigators have studied dysentery, bubonic plague, cholera, gas gangrene, influenza, tuberculosis, hemolytic streptococcal disease, encephalitis, primary atypical pneumonia, airborne infections, venereal diseases, infected wounds, burns, neurosurgery, X-rays, surgical sutures, shock, blood substitutes, treatment of gas casualties, convalescence and rehabilitation, insect and rodent control, antimalarial drugs, and the development and use of penicillin.

Among the most conspicuous achievements of this program are the following:

a. The acquisition, in civilian hospitals and laboratories, of sufficient knowledge of the therapeutic value of penicillin to warrant its official adoption by the medical divisions of the Army and Navy and to provide the impetus for the great production program that has made this remarkable drug available in large quantities for both military and civilian use.

b. Developments in insect repellents and insecticides, particularly DDT, important in guarding troops against insect-borne diseases such as typhus and malaria.

c. The study of human blood plasma which has led to use by the armed forces of serum albumin as a blood substitute, of immune globulins to combat infections, and of fibrin foam to stop bleeding.

d. The improvement and standardization of the treatment of malaria by atabrine.

e. The determination of the relative usefulness of sulfonamide drugs in the treatment of wounds and burns.

f. The physiological indoctrination of our airmen and the development of devices which enable them to endure the rigors of

53

high altitudes without disastrous loss of fighting capacity or life.

It is fair to say that without the Office of Scientific Research and Development or its equivalent few or none of the investigations listed above would have been carried out with the same speed and thoroughness. This research program to June 30, 1944, had cost over $15,000,000. Private funds were not available to finance this work.

3. Effect of War on Medical Research

Despite this imposing record of practical achievement, the war has seriously retarded the long-range development of medicine in ways perhaps not immediately apparent to the uninformed, but nevertheless with effects that will be longlasting. Because those physicians and scientists who have remained in their laboratories have, for patriotic reasons, devoted themselves to special problems raised by the exigencies of war, essential fundamental research has decreased to an extent which can only be viewed with grave concern.

Our hospitals and medical schools have suffered serious depletions of staff in order to supply the armed forces with needed physicians. Medical education has been hurried and impaired by the accelerated program, and the advanced training of young men has been in practically complete abeyance throughout the war. This diversion of physicians, coupled with an effective prohibition against graduate training in the ancillary sciences has left the fields of medical science barren and without the seed to produce a new generation of investigators. It will be many years before medicine fully recovers.

4. The Need for Continued Medical Research

It must be emphasized that nearly all that was good or apparently new in war medicine had its roots in civilian medicine. The pressure of war served chiefly to accelerate the development and large scale application to military needs of previously known facts. Medicine must consider now how to attack the medical problems of peace.

As President Roosevelt noted, the annual deaths in this country from one or two diseases alone are far in excess of the total number of lives lost by us in battle. This is true even though notable progress has been made in civilian medicine during the past three decades. Diabetes has been brought under control by the discovery of insulin; pernicious anemia by the use of liver therapy; and the once widespread deficiency diseases have been almost eradicated, even in the poorest income groups, by the discovery of accessory food factors and the improvement of the diet. Notable advances have been made in the early diagnosis of cancer, and in the surgical and radiation treatment of this dreaded disease.

In the period of 1900 to 1942, the average life expectancy of the American people increased from 49 to 65 years, largely as a result of the reduction in the death rates of infants and children. In the last two decades, the death rate from diseases of childhood has been reduced 87 percent. Deaths from scarlet fever have been reduced 92 percent, from whooping cough 74 percent, and from measles 91 percent. The death rate from diphtheria among children (5 to 14) is only one eighteenth what it was two decades ago. Only one-fourth as many children die today from tuberculosis and pneumonia as would if the mortality rate of 20 years ago still prevailed. The death rate among children between the ages of 5 and 14 from all causes combined was cut 57 percent between 1922 and 1942. For every three children who die under current conditions, more than seven would have died if the death rate of two decades ago had continued.

This reduction in the death rate in childhood has shifted the emphasis in medicine to the middle- and old-age groups, and particularly to the malignant diseases and the degenerative processes which are prominent in the later decades of life. Cardiovascular disease, including chronic disease of the kidneys, arteriosclerosis, and cerebral hemorrhage, now accounts for 45 percent of the deaths in the United States. Second in importance are the infectious diseases, and third is cancer. Added to these are many maladies (for example, the common cold, arthritis, asthma and hay fever, peptic ulcer) which, though infrequently fatal, cause incalculable disability.

Another aspect of the changing emphasis in clinical medicine is the increasing incidence of mental disease. Approximately 7,000,000 persons in the United States are mentally ill. More than one-third of the hospital beds in this country are filled with such persons at a cost of $175,000,000 annually. Each year nearly 125,000 mentally ill new patients are hospitalized.

In short, despite notable progress in prolonging the span of human life and in alleviating suffering, adequate methods of prevention and cure are not yet available for many diseases. Additional hospitals, physicians, and mechanisms for dispersing knowledge, however useful, cannot supply a complete solution. We simply do not know enough, and increased facilities for medical care will not supply the missing answers. The basic task faced by medicine is continued exploration of the human organism and the nature of disease. This exploration has only begun.

5. Importance of Fundamental Research to the Progress of Medicine

Research in medicine may be carried out effectively in two ways: First, by a coordinated attack on a particular disease; or second, by independent studies of the fundamental nature of the human body and its physiological mechanisms, of the nature of bacteria, viruses, and other agents of disease, and of the influence of environment on both. An example of the first method is the attack on malaria carried out under the Army, Navy, Public Health Service, the National Research Council, and the Office of Scientific Research and Development. The discovery of penicillin is an example of the second method: Fleming noted that a common mold, *Penicillium notatum*, inhibited the growth of a culture of bacteria in which it appeared as a contaminant. Thus an incidental observation in the course of studies unrelated to chemotherapy furnished the basis for the ultimate development of the most valuable chemotherapeutic agent known.

Discoveries in medicine have often come from the most remote and unexpected fields of science in the past; and it is probable that this will be equally true in the future. It is not unlikely that significant progress in the treatment of cardiovascular disease, kidney disease, cancer. and other refractory conditions will be made, perhaps unexpectedly, as the result of fundamental discoveries in fields unrelated to these diseases.

To discover is to "obtain for the first time sight or knowledge of some fact or principle hitherto unknown." Discovery cannot be achieved by directive. Further progress requires that the entire field of medicine and the underlying sciences of chemistry, physics, anatomy, biochemistry, physiology, pharmacology, bacteriology, pathology, parasitology, etc., be developed impartially.

6. The Place of Medical Schools and Universities in Medical Research

The medical schools and universities of this country can contribute to medical progress by carrying on research to the limit of available facilities and personnel, and by training competent investigators for an enlarged program in the future.

In some cases coordinated direct attacks will be made on special problems by teams of investigators from the medical schools, supplementing similar direct attacks carried on by the Army, Navy, Public Health Service, and other organizations. However, the main obligation of the medical schools and universities, in addition to teaching, will be to continue the traditional function of these institutions—that of providing the individual worker with an opportunity for the voluntary and untrammeled study in the directions and by the methods suggested by his imagination and curiosity. The entire history of science bears testimony to the supreme importance of affording the prepared mind complete freedom for the exercise of initiative. The special

duty and privilege of the medical schools and universities is to foster medical research in this way, and this duty cannot be shifted to Government agencies, industrial organizations, or any other institutions.

Because of their close relationship to teaching hospitals, the medical schools are in a unique position to integrate clinical investigation with the work of the departments of preclinical science, and to impart new knowledge to physicians in training. Conversely, the teaching hospitals are especially well organized to carry on medical research because of their close relationship to the schools, on which they depend for staff and supervision.

Not all our medical schools are equally developed. Because of inadequate financial support or lack of trained personnel, some of them can contribute little to medical research. A great increase in the resources of the Nation would be achieved by stimulating research in these less favored schools. It is imperative that we employ all possible methods of improving the research facilities and research staffs of our present medical schools before considering the establishment of new institutions.

7. Medical Research Under State Sponsorship in Great Britain

Although Federal aid for medical research was brought about in the United States largely under pressure of war. Government support of research has been general in Europe for many years. As a rule this support has been delegated to organizations separate from the ordinary Government bureaus in order to remove it as far as possible from political influence and to place the administration of funds in the hands of men experienced in research.

In Great Britain as early as 1911 the promotion of medical research was explicitly recognized as a responsibility of the State by the establishment of the Medical Research Committee, which became the Medical Research Council in 1920. The Council has administrative autonomy with general responsibility to a committee of ministers in the Privy Council. It receives money from both Parliament and nongovernmental sources specifically for furthering medical research and has no connection with any system of medical care or health insurance.

The Medical Research Council has continued to play an increasingly important and eminently successful role in its field. Through it Government support for medical research and the aid of medical science to the Government are assured.

Medical research in Great Britain also receives indirect Government aid through the University Grants Committee, a Standing Committee of the Treasury. Its members are independent experts of acknowledged repute and thoroughly familiar with the problems of university administration. The Committee's terms of reference are "To inquire into the financial needs of university education in the United Kingdom, and to advise the Government as to the application of any grants that may be made by Parliament toward meeting them."

Although the University Grants Committee does not give direct grants for specific medical research projects, it holds that research is one of the primary functions of a university and an indispensable element in the work of university teachers. Grants to the institutions are in the form of unrestricted funds with no portion earmarked for a specific purpose. Through a recent act of Parliament whereby this Committee is enabled to award $4,000,000 annually to medical schools and $2,000,000 to teaching hospitals, this indirect support of medical research by the Government has been substantially increased.

8. The Need for Federal Aid to Medical Research

Between World War I and World War II the United States overtook the other nations in medical research and forged ahead to a position of world leadership. If this leadership is to be maintained, some form of Government financial aid to the medical schools will be necessary. This view is accepted by the Committee and by nearly all whom the Committee has consulted.

Dr. A. N. Richards, Chairman of the Committee on Medical Research, reported to the Subcommittee on Wartime Health and Education of the Committee on Education and Labor of the United States Senate that, in connection with medical research, "The experience of the Office of Scientific Research and Development has proved that none of the universities which were called upon for Office of Scientific Research and Development work could afford to undertake it on the scale which the emergency demanded at the expense of its own resources. Hence, if the concerted efforts of medical investigators which have yielded so much of value during the war are to be continued on any comparable scale during the peace, the conclusion is inescapable that they must be supported by government."

At the same hearing, Dr. Lewis H. Weed, Chairman of the Division of Medical Sciences of the National Research Council, stated "* * * Much of medical research will necessarily have to be abandoned in the private and semiprivate institutions of the country unless Government subsidy is made available in some form for the general support of medical research."

Without Federal support American medical research will not stop, but without it our opportunities to advance medical knowledge cannot fully be exploited, and our objectives will be reached more slowly.

It has been computed that the annual budgets of the 77 medical schools in the United States total about $26,000,000. The portion of this sum spent for medical research cannot be determined accurately. Income from tuition amounts to $8,000,000, leaving a deficit of $18,000,000 annually. To meet this deficit the schools, apart from those connected with State universities and financed by the respective States, draw upon many sources.

A substantial part comes from university endowment, but during the past 10 years the amount of new endowment to medical schools has greatly diminished. At the same time the income from present endowment has been cut by one-third. With continued high taxation it is improbable that large gifts and bequests for scientific work can be expected in the future.

In many instances funds are allocated to the medical schools from tuition fees derived from other departments of the university.

Another source of research funds is the foundations, but, as in the case of the universities, the income from foundation endowment is decreasing.

Moreover, the foundations in general favor short-term grants to projects which carry promise of yielding immediate results.

Industry is a potential source of funds, but gifts from this source are usually for specific problems of a developmental nature. University alumni associations contribute only relatively small sums. Direct gifts from individuals are a substantial aid at times, but the medical schools must compete with all charities and churches for these funds. Furthermore, it is estimated that gifts from individuals, while perhaps more numerous, are far smaller in total than the large contributions of individual donors in the past.

When the funds available to a medical school are cut, the institution usually retrenches by curtailing the portion used for research. Overhead and teaching expenses must be met, and research becomes a luxury.

Finally, while research funds are decreasing, the costs of research are steadily rising. More elaborate and expensive equipment is required, supplies are more costly, and the wages of assistants are higher.

9. How Financial Aid Should be Supplied

Federal financial aid to the medical schools should be provided in three forms: General research funds, fellowships, and grants-in-aid.

General research funds

It is the Committee's opinion that unrestricted grants, with no portion earmarked for specific purposes, and with administration delegated to local research boards, would be the most valuable and productive form in which Government support could be given.

A medical school consists of a dozen or more semi-autonomous departments, each with its own budget. In the schools favored with a large endowment, research projects are constantly in progress in all departments; in financially weaker schools, the budget of a department may be too small to supply as much as a secretary for the department head, and research is, of course, a financial impossibility. Even in the most favored departments, the quality and quantity of research would be greatly increased if it were possible to employ an extra technical assistant or two, to purchase additional supplies or a necessary piece of equipment, to improve or enlarge animal quarters, or to meet other countless small financial requirements that may arise suddenly and may be of a temporary nature. In departments with small budgets such requirements are even more pressing. Many medical schools at present have small likelihood of securing grants-in-aid because they have neither personnel nor equipment to conduct successfully the type of research project appropriately financed by this method.

If a central agency were to attempt to meet item by item these many requirements by means of specific grants, the administrative costs would be prohibitive. The amount needed for each item is small, but the total amount needed by an institution may be relatively large.

Furthermore, a central agency would lack the flexibility to meet the rapidly varying and often temporary research needs that arise in the medical schools. A promising lead in research may prove patently false within a month or two. It is equally important that the project should then be stopped, and its personnel and equipment promptly diverted to more productive work, as it is that the project should have been given a trial.

A special use for general research funds would be to provide "junior fellowships" which would allow a medical student to interrupt his course, usually between the preclinical and clinical years, and to devote himself full-time to research for a year or two. The chances in this country for medical students to gain research experience prior to graduation are few, and as a result much research ability goes undiscovered. Candidates for these fellowships would be unknown to a central agency, which would have to rely entirely upon the judgment of the local research board for their selection. Hence it would be proper and economical to provide these fellowships from the general research funds administered by the local board.

The provision of funds as block grants to local research boards would exercise to greatest advantage the principle of decentralization of control of research; would eliminate costly overhead; would create a flexible mechanism to meet rapidly varying needs; would allow full play to the wisdom and experience of medical school faculties and administrators, whose knowledge in aggregate and whose particular knowledge of local needs must always exceed that of a central agency; would promote research in laboratories where it is now poorly developed; would foster investigations of an exploratory nature; and would provide the greatest and most effective stimulus to medical research.

Fellowships

Federal funds should be used to support fellowships, extending over periods up to 6 years, to be awarded by the Government agency to enable selected men to obtain training in research, to learn techniques in fields other than those of their basic scientific education, or to undertake research on a full-time basis. Since 1921 the fellowship program, supported by the Rockefeller Foundation and administered by the Medical Fellowship Board of the National Research Council, has made an important contribution to the advance of medical science and to the training of teachers and investigators in the United States. An increase in the number of such fellowships is greatly needed.

Grants-in-aid

A limited number of important research projects both of immediate and long-range consequence, will require special grants-in-aid. On occasion, through grants-in-aid, support should be given to medical schools, hospitals, or nonprofit scientific institutions to enable a senior investigator to develop the problems of his interest more rapidly and effectively.

10. *Estimated Cost of Program*

No final statement on costs is possible at this time. From information received from the deans of medical schools, from the expenditures of the Committee on Medical Research, and from other sources, it is estimated that approximately 5 to 7 million dollars annually can be used effectively in the immediate postwar period. A larger sum may be required when the program is fully underway. This estimate does not include the possible assumption of present commitments of the Office of Scientific Research and Development. A more definite statement would require prolonged study.

11. *The Need for an Independent Agency*

Advances in medical science have come and will continue to come preponderantly from medical schools or science departments of universities. Therefore the problem of improving medical research and of training more top-flight investigators is primarily one of aiding the medical schools and universities to utilize their research and educational facilities to the fullest extent.

In the Committee's opinion, medical research could best be promoted by the creation of an independent Federal agency.

This new organization would not conflict with the medical interests of existing Government agencies, none of which is primarily concerned with developing the basic medical sciences or with training personnel, both of which are functions of the universities. Some duplication of investigation would occur in problems in which civilian investigators and one or more Government agencies were mutually interested. However, it cannot be too strongly emphasized that, far from being wasteful, duplication is imperative in medical research, where each new discovery can be accepted only after repeated confirmation by independent observers approaching the problem from different points of view. The duplication is more apparent than real, as the results of independent investigators working on a common problem rarely agree exactly, and the differences are frequently the basis for new discoveries.

Rather than conflicting with existing agencies, the proposed body would supplement the research activities of these agencies in a valuable manner. Only through the efforts of such a body can our Government agencies be supplied with the necessary increase in numbers of expert personnel and with the all-important increase in basic scientific knowledge on which medical advance depends.

As the function of the proposed agency is broadly conceived, as it must be concerned not only with research but with the training of personnel required by all existing agencies, and as it must operate through non-Governmental education institutions, the future of which rests heavily upon private endowment or support by the States, it is the Committee's conviction that the Federal agency concerned with medical research should be created *de novo* and be independent of all existing agencies, none of which is sufficiently free of specialization of interest to warrant assigning to it the sponsorship of a program so broad and so intimately related to civilian institutions.

12. *Compensation*

The Committee believes that better effort will be put into the work of the agency by members if they are paid. The question of adjustment of salary from parent institutions should be left to the parties concerned.

It is estimated that members of the board of trustees and technical board, as proposed below, will be called upon to give an average of one-third of their time to the work of the agency. One-half the time of the aides may be required. This includes time devoted by members to the work of the agency at their official stations and in traveling.

Over the past 25 years there has been an increasing draft of expert personnel from the medical schools to meet the demand for scientists in activities related to the national welfare, until at present, even discounting the increased demands of war, many teachers and investigators are unable to discharge their responsibilities to the institutions which pay their salaries. A further increase in this borrowing of personnel without compensation can inflict only injury upon the medical schools.

Moreover, many competent investigators in medicine and surgery draw a negligible fraction of their income as salary, depending financially upon clinical practice. Participation in the work of the agency may interrupt this practice and the resulting loss of income may exclude such persons from service.

13. *Patent Rights*

The practice in regard to patent rights on discoveries and inventions bearing on human health varies in different medical institutions in this country. The Committee has made no effort to codify them, or to arrive at a generally acceptable policy.

It seems to the Committee that under the present patent laws the principle of patenting certain types of discoveries and inventions to exclude misuse is sound. Since perhaps the majority of institutions do not capitalize their patent privileges, and since such practice would be incompatible with Government sponsored research, it is suggested that, where a patent be granted on research which has been sponsored by Government in whole or in part, the ownership of the patent remain in the inventor, and that the Government receive, in addition to a royalty-free license, the power to require the licensing of others.

Part Two

Fundamental Principles Governing the Use of Federal Funds for Medical Research

As stated above, the Committee is convinced that Federal aid is necessary to ensure maximal progress in the development of medical science. It is also convinced that this aid, if misdirected, may do serious harm. It believes that among the major principles which should govern the application of Federal aid to medical research are the following:

a. Until experience has indicated the best plan of organization and procedure, the Federal agency created to aid medical research should be kept as flexible as possible. One of our colleagues has written "The common history of social organizations has been their creation in response to an idea, their flowering under the influence of the idea, their loss of the idea, and their perpetuation for the maintenance of the prestige of the office-holder." Only if authority to experiment with organization is written into its charter will an agency designed to aid medical research escape this fate.

b. The administration of Federal aid to medical research must be free from political influence and protected against special pressures.

c. Men who are experienced in research and who understand the problems of the investigator should administer the agency and determine its policies. Since the agency will be concerned primarily with basic scientific research in, and scientific training and policies pertinent to, endowed or State supported civilian institutions, and since the armed forces, the Public Health Service, and other existing governmental services have specialized interests, the Committee believes that it is as improper for any one of these services to hold the power of vote in matters pertaining to the proposed new agency as it would be for one or more members of the agency to vote in the medical councils of the services.

d. The agency should not attempt to dominate or regiment medical research but should function by creating greater opportunities and more freedom for investigation, and by aiding in cooperative efforts. It should not attempt to influence the selection of personnel, the conditions of tenure, the salary level, or other internal affairs of the institutions to which it gives aid.

e. Any program of Federal aid to medical research should be modestly initiated in terms of actual needs and conservatively increased as the capacity of the medical schools to utilize additional funds is demonstrated. If the Government spends too much in medical research, other funds will be driven out and the Government will be the sole source of support. The schools should remain free to elect the potential donor to whom they wish to apply. As Senator Pepper has stated, "Government can not, and must not, take the place of philanthropy and industry in the sponsorship of research."

f. The establishment of life-time research professorships, or of protracted research fellowships, at the expense of Federal funds is considered unwise. In exceptional instances, as for example when an investigator demonstrates unusual ability, or it is desirable to relieve a senior and experienced person from academic or clinical responsibilities in order to free him for research, support should be obtained from general research funds or through a grant-in-aid.

g. A grave danger in any effort to accelerate discovery is the ease with which the quality of the work can be lowered by encouraging men to undertake research who are inadequately prepared or unfitted for the task. Mediocre research work in medicine is not only apt to be useless, but may prove dangerous by misleading medical practice and by fostering false hopes in the public. This danger must be guarded against by constantly encouraging confirmatory work or "challenging investigations."

h. The agency should not serve merely as a mechanism for disbursing funds for particular research projects, but should always attempt to maintain a broad view of the needs of the whole field of medical research.

i. It is believed that it would be unwise for a national body concerned with medical research to give prizes or otherwise to dispense praise or blame. It is also believed that this agency should avoid even the semblance of scientific authority. What is acceptable or unacceptable in medicine must be established by tested methods of examination and not be made to appear as such because of the imprimatur of a national body.

j. The agency should come to share in the leadership of medical investigation by encouraging individual initiative and freedom of research, and with a careful avoidance of coercion and regimentation, which might lead not only to mediocre work but to disastrous impairment of the spirit of cooperation, and of research itself. Individual scientific curiosity, community of interest and regard for the common weal must in peace replace as a cohesive force the patriotism of war.

Part Three

Recommendations Outlining the Establishment of a "National Foundation for Medical Research" as an Independent Federal Agency

It is recommended that an independent agency of the Federal Government be established, to be known as the National Foundation for Medical Research.[1]

1. Composition of the Foundation

The foundation is to be composed of (a) a board of trustees, (b) a technical board, and (c) an executive secretary's office.

Board of Trustees

The board of trustees is to consist of five persons appointed without regard to the Civil Service Laws by the President of the United States and subject to confirmation by the Senate. They are to be chosen on the basis of scientific achievement and leadership, wide knowledge of medical problems, capacity for administration and organization, and with reasonable regard for geographical representation. The board of trustees is to elect its own chairman.

A member of the board of trustees is to serve on a part-time basis for a term of 5 years and is not to be eligible for reappointment. A member appointed to a vacancy caused by death or resignation is eligible for reappointment for a full term pro-

viding his short term has been less than 2 years. No two members serving simultaneously shall be chosen from the same institution. The successor to a retiring member shall not be chosen from the same institution except in unusual instances.

The original members of the board of trustees are to be appointed for 2, 3, 4, 5, and 6 years, respectively, in order to assure continuity and rotation. Whenever a vacancy occurs or is to occur, the chairman is to transmit to the President of the United States for his information a list of suitable candidates. In preparing this list, the chairman is instructed to seek the advice of the President of the National Academy of Sciences.

The chairman is to represent the Foundation in matters affecting medical research where the interests of other Government agencies are involved.

The board of trustees is to meet not less than once each month. At least one regular meeting each year is to be held in each of the following geographical areas: North Atlantic, South Atlantic, North Central, South Central, Rocky Mountain, and Pacific coast areas.

The board of trustees is to determine the broad policies of the Foundation. It is to appoint members of the technical board and is to have the authority to approve or disapprove

---

[1] Wherever used the term "medical research" is intended to include related aspects of dentistry, veterinary medicine, biology, entomology, protozoology, and similar fields.

all recommendations of the technical board. It may request the chairman and other members of the technical board to sit with it whenever necessary.

The board of trustees is to establish necessary liaison offices to insure a free exchange of information with all domestic and foreign agencies or services interested in medical research. It is to invite the Surgeons General of the Army, the Navy, the United States Public Health Service, the Air Force, or responsible officers of other domestic or foreign organizations as may be indicated, to appoint appropriate liaison officers to sit with it during deliberations of interest to those agencies. Liaison officers are not to have the power of vote.

Remuneration

Each member of the board of trustees is to be paid a salary of seventy-five hundred dollars ($7,500) per annum for that portion of his time which he devotes to the services of the Foundation. In accordance with Government regulations, a member is to receive travel expenses and suitable per diem to cover other costs when traveling.

Technical Board

A technical board, composed initially of 12 persons, is to be appointed, without regard to the Civil Service Laws, by the board of trustees. The members of the technical board are to be chosen on the basis of their knowledge and experience in special fields of medical research and the related sciences, and with reasonable regard for geographical representation. The office of a board member is to remain in his parent institution. At the discretion of the

board of trustees the membership of the technical board may be increased or decreased in number.

A member of the technical board should not ordinarily be considered eligible for reappointment, but a retired member may be appointed to the board of trustees. A member appointed to fill an unexpired term is eligible for reappointment for a full term. No two members serving simultaneously shall be chosen from the same institution. The successor to a retiring member shall not be chosen from the same institution except in unusual instances.

The chairman of the technical board is to be designated by the board of trustees. He is to represent the technical board before the trustees, is to call meetings of the technical board as frequently as necessary, and is to be responsible for the supervision of the activities of the board and the preparation of reports required by the board of trustees.

The original members of the technical board are to be appointed in groups of 4 to serve 2, 3, and 4 years, respectively, in order to assure continuity and rotation.

Remuneration

Each member of the technical board is to serve on a part-time basis for 3 years, and is to receive a salary to be determined by the Board of Trustees, but not to exceed five thousand dollars ($5,000) a year for that portion of his time which he devotes to the services of the Foundation. In accordance with Government regulations, a member is to receive travel expenses and suitable per diem to cover other costs when traveling.

Aides

Each member of the technical board may, with the approval of the technical board, appoint one or more aides without regard to the Civil Service Laws. These aides are to be selected on the basis of qualification in a special research field and are to serve on a part-time basis for periods up to 3 years.

As determined by the board of trustees, aides are to be compensated for time spent in the work of the Foundation, and when traveling are to receive travel expenses in accordance with Government regulations and a suitable per diem to cover other costs.

As aides are scientists in a potentially productive period, provision is to be made to insure that they remain professionally active, and that service with the technical board does not jeopardize their academic careers.

Committees

The technical board is to appoint ad hoc committees to advise with a particular member on medical problems. Members of such committees are to be appointed consultants with per diem compensation up to twenty-five dollars ($25), and in accordance with Government regulations are to receive travel expenses and suitable per diem to cover other costs when traveling.

Aides and committees appointed for a technical board member are to be discharged on the expiration of the member's term, but continued service may be invited by the member's successor.

Authority of the technical board

The technical board is to receive, review and recommend to the board of trustees on all requests for general research funds, fellowships, and grants-in-aid.

It is to take such steps as are necessary to put approved programs into effect.

It is to maintain reasonable supervision of work under general research funds and grants-in-aid and of the activities of Fellows, and keep the trustees informed on the progress of this work.

It is to arrange for the preparation of reports or appraisals as requested by the board of trustees.

Its members are to keep themselves informed on the status of pertinent medical problems, to which end they are authorized to convene round-table discussions, to invite competent persons to prepare summaries of specific problems, and to seek authoritative information in any other appropriate manner.

It is to receive and consider recommendations from individual investigators with regard to the further development of problems of possible scientific interest.

Office of the Executive Secretary

A full-time executive secretary is to be appointed by the board of trustees after consultation with appropriate Government fiscal and accounting agencies. The executive secretary is to organize administrative, fiscal, and accounting offices for the conduct of the business of the Foundation. Fiscal actions approved by the board of trustees are to be put into effect by the executive secretary and his affiliated officers.

Except for the executive secretary, all members of the staff of the executive secretary are to be drawn from qualified civil-service lists.

2. Functions of the Foundation

The functions of the Foundation are to be (a) to further medical research by providing financial aid through general research funds, fellowships and grants-in-aid; (b) to coordinate research in progress and to initiate new work considered essential; (c) to establish necessary liaison to secure a free exchange of medical information.

Financial Aid

a. *General research funds*

On application, a block grant may be made to a medical school for general use over a period of 1 to 10 years for the promotion of research provided the institution can present evidence that it can efficiently utilize for scientific research the funds requested, and that it is prepared to give a reasonable accounting of the expenditure of funds received. The institution is to have a research committee, drawn preferably from the executive faculty and active investigators, which is to be informed on all local research expenditures, and is to be responsible for the administration of the grant and for reports and accounting required by the Foundation.

The institutions are to be allowed wide latitude in the expenditure of general research funds, but these expenditures are to be subject to review periodically by the Foundation, which is to have the power of cancellation.

It is recommended that general research funds be used in part for junior fellowships to be awarded, without reference to the Foundation, to students working for an M.D. degree, in order to permit the recipients to devote 1 or 2 years on a full-time basis to acquiring more specialized knowledge of the techniques of medical research than is possible during the regular course. Junior fellowships are not to be used as scholarships to defray medical school tuition. The policy of each institution in regard to number of Junior Fellows, the value of the stipend, and other features of general importance is to be subject to review by the Foundation.

Formal discussions concerning renewal of general research funds should be completed 1 to 3 years in advance of termination.

If an application for general research funds is refused, the applicant institution may appeal directly to the board of trustees for a review.

In allocating general research funds, the Foundation is to consider both the immediate needs and promise of development of the applicant institutions, and is to take cognizance of the effects of such funds upon the support of medical schools by their parent institutions.

Equipment purchased under general research funds is to become the property of the institution to which the block grant is made.

b. *Fellowships*

Fellowships are to be awarded by the Foundation, for a period of 1 to 3 years, to approved applicants having the M.D., Ph.D., or D.D.S. degree or equivalent attainment, to enable the recipients to acquire research training, to undertake research, to learn special techniques, or to pursue studies in related sciences. Fellowships may be renewed for a period up to 3 years, but only in exceptional instances should the term of a fellowship exceed 6 years. The holder of a fellowship is to be publicly designated as a "Fellow of the National Foundation for Medical Research."

In the initial selection of Fellows, potentialities for development of leadership in medicine should be weighed as heavily as past performance in research work. Fellows are to be encouraged to take further work in the fundamental sciences to remedy any deficiencies in a contemplated research career, but fellowships are not to be used to provide residencies, or primarily for obtaining postgraduate degrees or for qualifying for Certification by the Specialty Boards.

Fellowships are primarily intended to enable men to receive research training and to engage in active research, but they should include experience in teaching or the clinical care of patients, as these exercises are essential to balanced research training and are imperative if a Fellow is to fit himself for. maximal usefulness in medicine or the medical sciences.

Fellowship stipends are to be determined by the Foundation with due consideration of university salaries paid persons with equivalent training and experience, and to the desirability of encouraging relatively senior men to devote themselves to research.

Research expenses of a Fellow may be met by the Foundation. If an institution matches insurance or annuity payments by its faculty, a similar payment is to be added by the Foundation to the Fellow's stipend.

A fellowship is to lapse automatically if a Fellow transfers to another institution without approval by the Foundation.

c. *Grants-in-aid*

On application, grants-in-aid extending for 1 to 10 years may be made to universities, medical schools, or other nonprofit scientific institutions for the support of specific projects or of specified investigators. Applications for grants-in-aid are to carry the endorsement of the applicant institution. Formal discussions

concerning renewal should be completed prior to the beginning of the last third of the period of the grant.

Reasonable overhead expenses may be included in the financial statement accompanying a request for a grant, but overhead payments are not to be automatic.

Reports are to be submitted under each grant as required by the Foundation. Equipment purchased under a grant is to become the property of the institution to which the grant is made.

If a request for a grant or for the extension of a grant is refused, the applicant institution may appeal directly to the board of trustees for a review.

Coordination and Initiation of Research

The Foundation is to consider methods designed to stimulate research, to improve research conditions in institutions where it is now not well developed, to effect coordination among investigators working in a common field, and to facilitate publication, dissemination, and experimental application of scientific information.

The Foundation is to initiate and support such new research work as may be indicated, but it is not itself to engage in research. Its integrative and catalytic efforts are to be carried out by recommendation and invitation rather than by direction.

3. *Reports*

The Foundation is to report annually to the President, in the form he requests, on the progress of work carried out under its authority. With the President's approval, all or part of the annual report is to be published.

4. *Authority to Modify Procedure*

The organization and responsibility of the Foundation are to be defined as broadly as possible. The authority to make and alter specific regulations and to experiment in procedures for fostering medical research is to be incorporated in the charter of the Foundation.

Report of the Committee on Science and the Public Welfare

Table of Contents

LETTER OF TRANSMITTAL

APRIL 16, 1945.

DR. VANNEVAR BUSH, *Director,*
Office of Scientific Research and Development,
16th and P Streets NW.,
Washington, D. C.

DEAR DR. BUSH:

It is with satisfaction that I hand you herewith a copy of the report of the Committee on Science and the Public Welfare.

We have had a number of meetings with good attendance and excellent discussion. We have unanimously agreed on practically all essential points. If the report aids in any degree in completing the task assigned you by the late President Roosevelt all members of the committee, I feel sure, will be gratified.

Sincerely yours,

ISAIAH BOWMAN, *Chairman,*
Committee on Science and the Public Welfare.

MEMBERS OF THE COMMITTEE

Isaiah Bowman, Chairman, President, Johns Hopkins University.
John T. Tate, Vice Chairman, Research Professor of Physics, University of Minnesota.
W. Rupert Maclaurin, Secretary, Professor of Economics, Massachusetts Institute of Technology.
Oliver E. Buckley, President, Bell Telephone Laboratories.
Walter C. Coffey, President, University of Minnesota.
Oscar S. Cox, Deputy Administrator, Foreign Economic Administration.
Bradley Dewey, President, Dewey & Almy Chemical Company.
Clarence A. Dykstra, Provost, University of California at Los Angeles.
Caryl P. Haskins, Director, Haskins Laboratories.
Edwin H. Land, President and Director of Research, Polaroid Corporation.
Charles E. MacQuigg, Dean of Engineering, Ohio State University.
Harold G. Moulton, President, Brookings Institution.
J. Hugh O'Donnell, President, Notre Dame University.
I. I. Rabi, Professor of Physics, Columbia University (recipient of Nobel Award).
Warren Weaver, Director for Natural Sciences, Rockefeller Foundation.
Robert E. Wilson, Chairman of the Board, Standard Oil Company of Indiana.
William E. Wrather, Director, U. S. Geological Survey, Department of Interior.

PREFACE

Dr. Isaiah Bowman was named by Dr. Vannevar Bush, Director of the Office of Scientific Research and Development, as chairman of a committee to consider this question raised by President Roosevelt in his letter of November 17, 1944, to Dr. Bush: "What can the Government do now and in the future to aid research activities by public and private organizations? The proper roles of public and of private research and their interrelationship should be carefully considered."

The Bowman Committee has confined its attention to research activities in the natural sciences, engineering, and agriculture. Clinical medicine has been considered by another committee. The support of the social sciences, it is believed, represents an important problem in itself which should be handled as a separate issue.

In analyzing the task assigned to the Bowman Committee, the project was divided into the following major questions:

1. What should the Government do to assist research in universities and nonprofit research institutes?

2. What should the Government do to assist scientific research conducted by the Government itself?

3. What should the Government do to assist research in industry?

4. What changes, if any, should be made in our present tax structure to stimulate industrial research?

5. What policy should the Government follow to encourage greater international interchange of scientific knowledge and engineering art after the war?

6. What are the proper roles of public and private research?

The Committee was divided into working groups to consider each of these questions except the last. The whole report is concerned with the basic problem of the proper roles of public and private agencies in scientific research. The analyses undertaken by the working groups have been combined into a final report which is submitted herewith.

In addition to numerous meetings of the subcommittees the main Committee has held three full meetings, the first of which was devoted to resolving the problem into its major parts, the second to discussing the concept of the Federal Government in relation to research, and the last to considering the recommendations of the subcommittees.

Interest in the question of Federal aid to research reflects widespread recognition by the American people that the security of a modern nation depends in a vital way upon scientific research and technological progress. It is equally clear that public health, higher standards of living, conservation of national resources, new manufacturing which creates new jobs and investment opportunities—in short, the prosperity, well-being and progress of the American Nation—all require the continued flow of new scientific knowledge. Even if a nation's manpower declines in relative numbers, even if its geographical frontiers become fixed, there always remains one inexhaustible national resource—creative scientific research.

In view of the importance of science to the Nation, the Federal Government, by virtue of its charge to provide for the common defense and general welfare, has the responsibility of encouraging and aiding scientific progress. It has recognized this responsibility by providing research laboratories within the structure of government, by providing a climate of law within which industry could progress on its own initiative, and by making limited appropriations to certain types of educational institutions. Study of the present status of research has shown convincingly that certain basic parts of our research structure require increased financial support. Since the evidence is clear that private sources cannot assume the entire burden, the committee has been forced to the conclusion that an increased measure of direct Federal aid is necessary. We believe that it is possible to devise methods whereby great benefits to research may be achieved by such aid without sacrificing the freedom essential for scientific advance or the academic independence of our traditional institutions.

We therefore urge that the Federal Government take a more active interest in promoting scientific research, and in assuring that the Nation gain therefrom the benefits of increased security and increased welfare. We are convinced that the most effective way for the Federal Government to serve these purposes is to provide to our educational institutions and research institutes support for basic research and training for research. By so doing, the Government will increase the flow of new knowledge and the supply of young scientists trained in research. It is on this new knowledge that applied science must build, and it is from the ranks of those trained in research that the leaders in applied science must come.

If this new knowledge and an adequate supply of trained men are provided, it is our opinion that the ordinary course of industrial activity can be relied upon to convert to practical application in industry most of the advances made in research. However, we believe that in certain instances measures can and should be devised to expedite the transition from scientific discovery to technological application. To this end we recommend that procedures be devised for supplying research information to small companies and stimulating them in the application of the latest technology.

In the international sphere the lack of any official Federal support for scientific meetings or experimental programs organized on an international scale has been a frequent source of embarrassment and difficulty. By providing official recognition and financial support to such undertakings the Government could do much to facilitate scientific interchange and promote international good will.

A National Research Foundation

We believe that our national and international needs and responsibilities in the field of science require the creation of a new Federal instrumentality. We therefore recommend that a National Research Foundation be created for the promotion of scientific research and of the applications of research to enhance the security and welfare of the Nation.

The control of the Foundation should be in the hands of a board of trustees. This board should be appointed by the President of the United States from a panel nominated by the National Academy of Sciences.

The Foundation shall be empowered, among other things, to:

1. Distribute funds in support of scientific research in educational and nonprofit research institutions, such research to be wholly under the control of such institutions.

2. Initiate and finance, in appropriate agencies, research projects for which existing facilities are unavailable or inadequate.

3. Establish scholarships and fellowships in the natural sciences.

4. Promote dissemination of scientific and technical information.

5. Support international cooperation in science by providing financial aid for international congresses, worldwide associations of scientific societies and scientific research programs organized on an international basis.

6. Devise methods of improving the transition between pure research and its practical applications in industry.

Research Carried on by the Federal Government

Research carried on directly by the Federal Government represents an important part of our total research activity and needs to be strengthened and expanded after the war. Expansion, however, should be limited to fields of inquiry and service which are of public importance and are not adequately carried on by private enterprise.

To increase the effectiveness of research done within the various departments and laboratories of Government a number of important changes in existing practices are desirable.

1. The most important single factor in scientific and technical work is the quality of personnel employed. Separate and distinct procedures for recruiting and classifying scientific personnel are warranted by the exacting technical requirements in these services. No one change from current practice would do more to improve the quality of research conducted by the Government than to establish a separate branch of the Civil Service for scientific and technical positions.

2. A general up-grading of positions and salaries in the scientific services of Government, accompanied by a careful selection of new talent, would be a major contribution to improvement of the quality of research conducted by the Government.

3. Research programs of Government should be assured in terms of their long-run objectives. Appropriations by Congress to the principal Government scientific departments should be made in lump sums for broad programs of research extending over several years. Appropriations within the assured sum might then be made available as at present in the annual budget.

4. A permanent science advisory board should be created to consult with Government agencies and to advise the executive and legislative branches of Government as to the policies and budgets of Government agencies engaged in scientific research.

Environmental Aids to Industrial Research

The structure of Federal taxation and the operation of the patent system have an important impact on the research and development policies of industry. In designing postwar taxes, consideration should be given to increasing incentives to industrial research. The proper treatment of research costs for tax purposes should receive clear legislative definition. Specific recommendations on this point are included in the main body of the report.

Part One

Introduction

President Roosevelt has asked:

What can the Government do now and in the future to aid research activities by public and private organizations? * * * The information, the techniques, and the research experience developed by the Office of Scientific Research and Development and by the thousands of scientists in the universities and in private industry, should be used in the days of peace ahead for the improvement of the national health, the creation of new enterprises bringing new jobs, and the betterment of the national standard of living. New frontiers of the mind are before us, and if they are pioneered with the same vision, boldness, and drive with which we have waged this war we can create a fuller and more fruitful employment and a fuller and more fruitful life.

The President's request reflects widespread recognition by the American people that the security of a modern nation depends in a vital way upon scientific research and technological progress. It is equally clear that public health, higher standards of living, conservation of national resources, new jobs and investment opportunities—in short, the prosperity, well-being and progress of the American Nation—all require the continued flow of new scientific knowledge. Even if a nation's manpower declines in relative numbers, even if its geographical frontiers become fixed, there always remains one inexhaustible national resource—creative scientific research.

The advanced state of technology in the American economy, of which we are justly proud, could not have been realized without sound institutional foundations. Our public and private universities and nonprofit research institutes, our industrial research laboratories, the research agencies operated by the State and Federal Governments, all constitute part of a cooperative pattern within which tremendous achievements have already been made. We are confident that within that same framework even greater developments in science will mark the future.

The continued progress of science is a matter of the highest national importance. The Federal Government, by virtue of its charge to provide for the common defense and general welfare, has the responsibility of encouraging and aiding such progress. It has recognized this responsibility in the past by providing research laboratories within the structure of government, by providing a climate of law within which industry could advance on its own initiative, and by making limited appropriations to certain types of educational and research institutions. As far as the committee can determine, there is no major dissent from the view that the first two methods of aiding scientific progress fall within the proper function of government.

The time has come, however, for a careful evaluation of the questions raised by direct Federal aid to private institutions. Our universities clearly stand in need of increased financial support if they are to strengthen their basic contributions to the scientific life of the Nation. Financial aid may also be required to speed up the transition between basic discoveries in university laboratories and their practical industrial applications. The committee has therefore felt compelled to examine from the standpoint of public policy the question: "Is a substantial increase in Federal financial aid to scientific research in educational and other nonprofit research institutions necessary and desirable?"

If the necessity were not clearly demonstrable, several considerations might argue for the undesirability of such Federal support. These center upon the fear that Federal aid might lead to centralized control. It is the firm conviction of the committee that centralized control of research by any small group of persons would be disastrous whether such persons were in government, in industry, or in the universities. There might be a danger, too, that increased Federal aid would discourage existing sources of financial support. Private individuals might lose interest in contributing to research institutions and the great foundations might turn their attention to other fields. The States might reduce the support given their large universities. These varied sources of support have contributed materially to the development of vigorous centers of independent initiative throughout the United States and prevented control by any one group.

The committee has had to weigh these considerations against an analysis of the adequacy of the over-all support for science in America relative to the needs of society. Our national pre-eminence in the fields of applied research and technology should not blind us to the truth that, with respect to pure research—the discovery of fundamental new knowledge and basic scientific principles —America has occupied a secondary place. Our spectacular development of the automobile, the airplane, and radio obscures the fact that they were all based on fundamental discoveries made in nineteenth-century Europe. From Europe also came formulation of most of the laws governing the transformation of energy, the physical and chemical structure of matter, the behavior of electricity, light, and magnetism. In recent years the United States has made progress in the field of pure science, but an examination of the relevant statistics suggests that our efforts in the field of applied science have increased much faster so that the proportion of pure to applied research continues to decrease.

Several reasons make it imperative to increase pure research at this stage in our history. First, the intellectual banks of continental Europe, from which we formerly borrowed, have become bankrupt through the ravages of war. No longer can we count upon those sources for fundamental science. Second, in this modern age, more than ever before, pure research is the pace-maker of technological progress. In the nineteenth century, Yankee mechanical ingenuity, building upon the basic discoveries of European science, could greatly advance the technical arts. Today the situation is different. Future progress will be most striking in those highly complex fields — electronics, aerodynamics, chemistry—which are based directly upon the foundations of modern science. In the next generation, technological advance and basic scientific discovery will be inseparable; a nation which borrows its basic knowledge will be hopelessly handicapped in the race for innovation. The other world powers, we know, intend to foster scientific research in the future. Moreover, it is part of our democratic creed to affirm the intrinsic cultural and aesthetic worth of man's attempt to advance the frontiers of knowledge and understanding. By that same creed the prestige of a nation is enhanced by its contributions—made in a spirit of friendly cooperation and competition—to the world-wide battle against ignorance, want, and disease.

The increasing need for the cultivation of science in this country is only too apparent. Are we equipped to meet it? Traditional support from private gifts, from endowment income, from grants by the large foundations, and from appropriations by State legislatures cannot meet the need. Research in the natural sciences and engineering is becoming increasingly costly; and the inflationary impact of the war is likely to heighten the financial burden of university research. The committee has considered whether industry could or should assume most of the burden of support of fundamental research or whether other adequate sources of private assistance are in sight. The answer appears to be in the negative.

The committee has therefore become convinced that an increased measure of Federal aid to scientific research is necessary. Means must be found for administering such aid without incurring centralized control or discouraging private support.

Basically this problem is but one example of a series of similar problems of government in a democracy. Many of our important political decisions involve the necessity of balancing irreducible national functions against the free play of individual initiative. It is the belief of this committee that if certain basic safeguards are observed in designing a plan for Government support to science, great benefits can be achieved without loss of initiative or freedom.

The experience of the land-grant colleges represents an important precedent. The scale of Federal aid has been modest but has led to very significant results especially in agriculture; it has not led to domination by small groups; it has not been capricious and uncertain. On the contrary, it has progressed on a slowly expanding scale for over 80 years. No evidence has been brought before the committee that this sort of Federal aid has discouraged other sources of support. The land-grant colleges are examples of harmonious cooperation among State and Federal Governments, private individuals, and industry. American experience with support of higher education by State and local governments has been extremely satisfactory, our vigorous State universities standing as impressive testimonials.

The committee foresees that an increased measure of Federal support will raise new problems. We have, therefore, carefully considered the possibility of increasing Federal aid for scientific research without, at the same time, introducing undesirable paternalism. For, in order to be fruitful, scientific research must be free — free from the influence of pressure groups, free from the necessity of producing immediate practical results, free from dictation by any central board.

Many have been impressed by the way in which certain fields of applied science have benefited, during the war, from an increased measure of planned coordination and direction. It has thus been very natural to suppose that peacetime research would benefit equally from the application of similar methods. There are, of course, types of scientific inquiry that require planning and coordination, and a large degree of control is inevitable and proper in applied research. However, there are several reasons why pure science in peacetime cannot wisely or usefully adopt some of the procedures that have worked so well during the war. War is an enterprise that lends itself almost ideally to planning and regimentation, because immediate ends are more rigidly prescribed than is possible in other human activities. Much of the success of science during the war is an unhealthy success, won by forcing applications of science to the disruption or complete displacement of that basic activity in pure science which is essential to continuing applications. Finally, and perhaps most important of all, scientists willingly suffer during war a degree of direction and control which they would find intolerable and stultifying in times of peace.

It is the belief of this committee that increased support of research in American universities and nonprofit institutes will provide the most positive aid to science and technology. But we do not believe that any program is better than no program— that an ill-devised distribution of Federal funds will aid the growth of science. Our concrete proposals seek to augment the *quality* as well as the quantity of scientific research. We believe that there are historical precedents of Government aid to research, both in this country and abroad, which show the possibility of providing, within the framework of sound administrative practice, sustained nonpolitical grants which would operate in such a manner as to call forth from existing institutions even greater initiative, effort, and accomplishment.

The organization or instrument finally set up should not attempt to play the role of an all-seeing, all-powerful planning board trying to guide in detail the normal growth-processes of science. The first and most essential requirement is that the groups administering a program of research assistance be composed of men of the highest integrity, ability, and experience, with a thorough understanding of the problems of science. The committee believes that an independent Government body, created by the Congress, free from hampering restrictions, staffed with the ablest personnel obtainable, and empowered to give sustained and far-sighted assistance to science with assurance of continuing support, would constitute the best possible solution.

It is our belief that the desired purposes can best be served and the possible dangers minimized by centering the responsibility for this program in a new organization, a National Research Foundation, whose function should be the promotion of scientific research and of the applications of research to enhance the security and welfare of the Nation.

Part Two

Present Status and Trends in American Science

To aid in formulating policies of assistance to research, it will be helpful first to analyze the important types of scientific activity and to sketch the development of the principal types of American scientific institutions.

A. The Nature of Scientific Research

Scientific research may be divided into the following broad categories: (1) pure research, (2) background research, and (3) applied research and development. The boundaries between them are by no means clear-cut and it is frequently difficult to assign a given investigation to any single category. On the other hand, typical instances are easily recognized, and study of them reveals that each category requires different institutional arrangements for maximum development.

1. *Pure Research*

Pure research is research without specific practical ends. It results in general knowledge and understanding of nature and its laws. This general knowledge provides the means of answering a large number of important practical problems, though it may not give a specific solution to any one of them. The pure scientist may not be at all interested in the practical applications of his work; yet the development of important new industries depends primarily on a continuing vigorous progress of pure science.

One of the peculiarities of pure science is the variety of paths which lead to productive advance. Many of the most important discoveries have come as a result of experiments undertaken with quite different purposes in mind. Statistically it is certain that important and highly useful discoveries will result from some fraction of the work undertaken; but the results of any one particular investigation cannot be predicted with accuracy.

The unpredictable nature of pure science makes desirable the provision of rather special circumstances for its pursuit. Pure research demands from its followers the freedom of mind to look at familiar facts from unfamiliar points of view. It does not always lend itself to organized efforts and is refractory to direction from above. In fact, nowhere else is the principle of freedom more important for significant achievement. It should be pointed out, however, that many branches of pure science increasingly involve the cooperative efforts of numerous. individuals, and expensive capital equipment shared by many workers.

By general consent the discoveries of pure science have for centuries been immediately consigned to the public domain and no valid precedent exists for restricting the advantages of knowledge of this sort to any individual, corporation, State, or Nation. All the people are the beneficiaries. Governments dedicated to the public welfare, therefore, have a responsibility for encouraging and supporting the production of new knowledge on the broadest possible basis. In the United States this responsibility has long been recognized.

2. Background Research

The preparation of accurate topographic and geologic maps, the collection of meteorological data, the determination of physical and chemical constants, the description of species of animals, plants, and minerals, the establishment of standards for hormones, drugs, and X-ray therapy; these and similar types of scientific work are here grouped together under the term background research. Such background knowledge provides essential data for advances in both pure and applied science. It is also widely used by the engineer, the physician and the public at large. In contrast to pure science, the objectives of this type of research and the methods to be used are reasonably clear before an investigation is undertaken. Thus, comprehensive programs may be mapped out and the work carried on by relatively large numbers of trained personnel as a coordinated effort.

Scientific work of this character is necessarily carried on in all types of research organizations — in universities, in industry, and in Government bureaus. Much of it evolves as a necessary byproduct either of applied research or of development. Only very rarely, however, does the knowledge obtained emerge in patentable form and the public welfare is usually best served by prompt publication of the results.

There seems to be little disagreement with the view that these surveys and descriptions of basic facts and the determination of standards are proper fields for Government action and that centralization of certain aspects of this work in Federal laboratories carries many advantages. There are few private organizations equipped to carry out more than a small fraction of the research needed in these fields. And it is obvious, for example, that topographic maps are most useful when maps for the entire country observe similar rules in regard to scale, contour lines, conventional markings for roads, dwellings, etc. Similarly, standard units for hormones should be based on uniform test procedures and be stated, so far as is possible, in uniform units. The Federal Government has recognized these responsibilities in principle and the Bureau of Standards serves as an excellent example of how such work can be carried out most efficiently.

Recent technical advance in such fields as synthetic chemistry and industrial biology have resulted in a stream of new compounds and materials too rapid for present laboratories to catalogue. Many substances of great potential usefulness are either completely unknown, or their properties inadequately described. Complex minerals such as coal, and a wealth of agricultural products, are composed of chemical compounds, any one of which may become the basis of a new industry. What is needed is enough knowledge about their potentialities to justify the private investment necessary for their practical application. If the problem is left entirely in private hands, progress may be very slow. At present, only the larger industrial laboratories have the capacity to engage extensively in such research. It seems desirable, therefore, for the Government to arrange for work of this sort, either in its own laboratories or in outside institutions, and to make the results of this research generally available in a systematic manner.

3. *Applied Research and Development*

Applied research and development differs in several important respects from pure science. Since the objective can often be definitely mapped out beforehand, the work lends itself to organized effort. If successful, the results of applied research are of a definitely practical or commercial value. The very heavy expenses of such work are, therefore, undertaken by private organizations only in the hope of ultimately recovering the funds invested.

In several fields, admittedly, such as agriculture and in various special industries where the individual producing units are small and widely dispersed, the presence of a profit motive does not ensure the existence of adequate research and development. The substantial research work initiated by the Department of Agriculture has developed in response to these special needs.

The distinction between applied and pure research is not a hard and fast one, and industrial scientists may tackle specific problems from broad fundamental viewpoints. But it is important to emphasize that there is a perverse law governing research: Under the pressure for immediate results, and unless deliberate policies are set up to guard against this, *applied research invariably drives out pure.*

The moral is clear: It is pure research which deserves and requires special protection and specially assured support.

B. Development of Scientific Research in the United States

During the colonial period of American history, scientific work was carried on in random, sporadic fashion, and for the most part outside the universities. Franklin and Jefferson are outstanding examples of the type of gifted amateur whose influence upon American science continued to be felt well into the nineteenth century. In the first decades of the Republic, the older American colleges began to give science increased attention in the curriculum. But despite the presence on their faculties of such outstanding individuals as the Sillimans, Louis Agassiz, and Joseph Henry, it cannot be concluded that the colleges were active centers of research, or that science received much emphasis in institutions which, if they were not so exclusively concerned with religious instruction as heretofore, were still devoted to the ideals of a liberal education along the lines of strict classical and literary tradition.

With the college environment inimical or at least cool toward the growth of scientific research, neither Government support nor private endowment was available in the United States for the promotion of pure research until late in the nineteenth century. This is in marked contrast to the principal European countries where, almost without exception, science was directly supported by the governments. Gradually in response to a steadily increasing need, the Federal Government established the scientific bureaus that it needed to fulfill its obligations to the public.

During the course of the century it created the Coast and Geodetic Survey, the Naval Observatory, the Department of Agriculture, and the Geological Survey. In 1836, to cite an early example of Federal support of a scientific venture, the Wilkes Exploring Expedition was authorized "to expand the bounds of science and to promote knowledge." But the practical nature of all these activities is evident. Despite several eloquent expressions by scientific men of the important long-run utility of sponsoring pure science, the Congress turned a deaf ear to all proposals for creating scientific institutions having anything but limited and strictly utilitarian purposes. Washington's plan for a national university, and the various suggestions for a Government-sponsored academy or a national institution had the support of public figures like Jefferson, Madison, and John Quincy Adams but were unpopular in Congress and were often strenuously opposed by the older private colleges.

If Government support for science was not forthcoming, neither was support from private gifts or bequests. It is significant that the first considerable sum for the support of pure science came from a foreigner, the Englishman James Smithson, with whose bequest Congress—after debating its acceptance and disposition for nearly 10 years — created the Smithsonian Institution.

As a result of the profound forces which were converting America in the last decade of the nineteenth century from an essentially backward agricultural Nation to a world power, changes took place in our attitudes toward science and learning and toward the encouragement that should properly be accorded them. The State universities and land-grant colleges grew and prospered through generous public support. Science also became one of the beneficiaries of the private fortunes built up in the later nineteenth century. Whereas earlier it had been evident that only the Government could assume the burden of erecting and supporting an astronomical observatory, there were now men like James Lick with fortunes large enough to build and endow such expensive centers of research. Equally important were the contributions of private philanthropy in developing universities and in the direct support of research through the creation of nonprofit science institutions and philanthropic foundations.

Two of our best-known privately endowed institutions devoted to pure research, the Carnegie Institution of Washington and the Rockefeller Institute, were created shortly after the turn of the century. From the same gigantic fortunes stemmed the Rockefeller Foundation and the Carnegie Corporation. Their tremendous contributions to the progress of scientific research, not only in America, but throughout the world, cannot be exaggerated.

The latter part of the nineteenth and the early twentieth centuries witnessed the development of the American medical schools, which today serve as research centers not only for applied or practical medicine but for fundamental research in many biological problems which are basic to medicine. The medical schools appear to have been particularly attractive objects of private philanthropy. Various factors, such as the regulation of standards by the profession at large and the active interest of two or three of the largest foundations, have given to the medical schools of the country a uniformly advanced status not enjoyed by other *divisions of our* universities. In fact, only in the case of medical schools can the United States be said to excel all other countries in the number of first-rate research institutions per unit of population.

Almost equally significant is the growth of the Federal Government's own scientific bureaus. The existing agencies and departments, especially the Department of Agriculture, underwent an extraordinary development. An outstanding feature was the expanding program of grants-in-aid to the State agricultural experiment stations. The first decades of the twentieth century saw the creation of a number of new scientific bureaus and laboratories: the Bureau of Mines, the Bureau of Standards, and the National Institute of Health. The First World War led to the creation of the principal service laboratories, the Naval Research Laboratory, for example, and the National Advisory Committee for Aeronautics. By 1932 the total Government expenditure for research had risen to over 40 million dollars, more than double the figure for 1922.

But no factor in the gradual emergence of American science from its dependent state is more striking than the growth of research laboratories in industry. Prior to 1880 there were few, if any, commercial laboratories worthy of the name; but in the last decades of the nineteenth century powerful new industries, especially in the electrical field, grew out of basic technological discoveries and the inventive genius of men like Bell, Edison, and Elihu Thomson. Firms in these new industries almost from the outset adopted the policy of maintaining their lead by energetic pro-grams of scientific and technological research resulting in patents based in large part on the work of their own laboratories.

The First World War provided a further stimulus to the growth of commercial laboratories by revealing the inadequacies of our position in industrial research as compared to Germany, especially in the chemical field. Much of our present chemical industry, together with its vast research potential, grew up in response to needs which were demonstrated in the war, aided by the availability of patents seized from their former German owners.

C. The National Research Budget

The over-all picture of the development of research in the United States, as reflected in the changed structure and magnitude of the national research expenditures of the last 15 years, is shown in table I and in the corresponding figure I.

Since statistical information is necessarily fragmentary and dependent upon arbitrary definitions, most of the estimates are subject to a very considerable margin of error. Nevertheless, the following generalizations seem warranted:

(1) Of the three principal groups engaged in research, private industry contributes by far the largest portion of the total national expenditures, with the Government coming next and the educational institutions last.

(2) Research expenditures of industry, Government, and industrial institutes have been expanding considerably more rapidly (fig. I), than research in universities and science institutes.

During the war, the Government expanded its research budget from

$69,000,000 in 1940 to $720,000,000 in 1944. Not all of this large increase took place in Government laboratories. Substantial sums went to industry and to the universities. This resulted in changing the trend of university research expenditures. The universities spent $28,000,000 on

Table I

Scientific Research Expenditures and National Income

Year	National income [1]	Industry [2]	Nonprofit industrial research institutes [3]	Government (Federal and State) [4]	Colleges and universities [5]	Research Institutes [6]	Total scientific research expenditures
	Millions	Thousands	Thousands	Thousands	Thousands	Thousands	Thousands
1920	$74,200	$29,468	--------	--------	--------	--------	--------
1921	59,400	37,400	--------	--------	--------	--------	--------
1922	60,700	44,000	--------	--------	--------	--------	--------
1923	71,600	50,000	--------	$15,615	--------	--------	--------
1924	72,100	58,000	--------	16,336	--------	--------	--------
1925	76,000	64,000	--------	18,087	--------	--------	--------
1926	81,600	70,000	--------	16,995	--------	--------	--------
1927	80,100	75,928	--------	17,119	--------	--------	--------
1928	81,700	88,000	--------	17,757	--------	--------	--------
1929	87,200	106,000	--------	22,825			
1930	77,300	116,000	$560	24,066	$20,353	$5,212	$166,191
1931	60,300	131,320	1,240	26,945	--------	5,218	--------
1932	42,900	120,000	990	40,081	24,840	5,159	191,070
1933	42,200	110,268	740	--------	--------	4,887	--------
1934	49,500	124,000	1,540	22,243	19,286	4,767	171,836
1935	54,400	136,000	2,470	25,328	--------	4,785	--------
1936	62,900	152,000	2,530	33,891	25,000	4,701	218,122
1937	70,500	160,000	3,580	40,786	--------	4,635	--------
1938	64,600	177,168	4,080	49,382	28,496	4,596	263,722
1939	70,829	200,000	5,000	--------	--------	4,531	--------
1940	77,809	234,000	6,110	69,136	31,450	4,549	345,245
1941	96,900	--------	9,139	207,259	--------	--------	--------
1942	122,200	--------	14,079	332,151	39,575	--------	--------
1943	149,400	--------	--------	561,507	--------	--------	--------
1944	160,700	--------	--------	719,813	--------	--------	--------

[1] Kuznets, Simon S., *National Income and Its Composition, 1919-38*, Vol. I (New York, National Bureau of Economic Research, 1941), p. 137.

[2] National Resources Committee, *Research—A National Resource*, Vol. II, *Industrial Research* (Washington, Supt. Docs., 1938), p. 174; Perazich, G. and Field, P., *Industrial Research and Changing Technology* (Philadelphia, WPA, National Research Project, Rep. No. M-4, Jan. 1940), p. 65.

[3] Includes the industrial research institutes supported primarily by contributions from industry. Estimated $5,000,000 spent by nonprofit industrial research institutes for 1939 and extrapolated for other years by the Battelle Memorial Institute figures given in their publication *Research in Action* (Columbus, 1944), p. 51.

[4] Report on Federal Government expenditures on scientific research. Excludes Federal grants to agricultural experiment stations. 3 percent of Federal Government expenditures estimated as equivalent to scientific research expenditures by the States, exclusive of their grants to agricultural experiment stations and colleges and universities, which are included in the expenditures by the latter. 1940-44 Federal Government figures do not include grants to "educational institutions and foundations."

[5] The National Resources Committee reported that $50,000,000 were spent on research by all colleges and universities in 1935-36. Based on the surveys by the Bowman Committee, it was estimated that $25,000,000 of this were for expenditures on research in the natural sciences. The trend shown in research expenditures of a large sample of universities and colleges was used to extrapolate for years other than 1936. Figures include grants from foundations and from the Government for agricultural experiment stations.

[6] Includes the endowed research institutes which are not connected with any industry nor an integral part of any university, such as the Rockefeller Institute of Medical Research, the Wistar Institute, the Carnegie Institution of Washington, the Marine Biological Laboratory at Woods Hole, etc. The estimates have been made upon published information and questionnaires. The trend shown in the institutions on which complete information was available was used to extrapolate the research expenditures in other research institutes. It was estimated that six institutes constituted 75 percent of the total expenditures.

86

research in 1938, while in 1943–44 the Office of Scientific Research and Development, alone, contracted for $90,000,000 of research in universities and colleges.

Certain problems which should be considered in planning for a national postwar research program, and some guidance in meeting these problems, can be ascertained from a study of the basic prewar trends and relationships. Particularly important is a study of the relative expenditures for pure and applied research. According to the best available estimates, industry before the war devoted about 5 percent of its research budget, or $9,000,000,[1] and Government about 15 percent, or $7,500,000, to pure research. Colleges, universities, and endowed research institutes spent 70 percent of their research budgets, or a total sum of nearly $23,000,000 in this way.

Total national expenditure for pure science thus amounted to approximately $40,000,000 while that for applied reached a figure of $227,000,000, a ratio of nearly 1 to 6. In England, where the development of industrial research is, admittedly, very retarded, the corresponding ratio of pure to applied research is estimated at 1 to 1.2.[2]

In the decade from 1930 to 1940 applied research was expanding much more rapidly in the United States than was pure research. During this period industrial research expanded by 100 percent and governmental research by 200 percent. Research in colleges and universities increased by 50 percent, and the endowed research institutes (which were primarily devoted to pure research) declined by nearly 15 percent. It may be concluded, therefore, that since governmental and industrial expenditure is growing so rapidly, relative to that of the universities, generous support to university research is essential if the proportion of pure to applied research is to be maintained at anything like the previous relationship.

This support will have to include substantial expenditures for capital facilities. The great decline in capital outlays of privately supported institutions is very striking.

[1] In the year 1938.

[2] Computed from research budgets listed by Bernal, J. D., *The Social Function of Science* (London, Routledge, 1939).

Annual Expenditures for Capital Outlay
All Institutions of Higher Education
Millions of Dollars

Fiscal year	Privately supported institutions	Publicly supported institutions
1929–30	$73.1	$36.6
1931–32	56.8	35.0
1933–34	18.1	11.4
1935–36	15.3	32.1
1937–38	29.6	40.9
1939–40	20.6	63.6
1941–42	19.8	31.8

Source: *Biennial Surveys of Education* (Washington, U. S. Office of Education).

Some portion of the new plant and equipment constructed during the last few years for the purpose of war research can be converted to peacetime uses. Nevertheless, a considerable amount of new investment will have to be undertaken after the war. There can be no doubt that such new construction could constitute one of the most productive kinds of public and private investment.

Next to the achievement of an adequate total volume of research activities and the establishment of a proper proportion between its pure and applied phases, maintenance of a continuous and steady expansion should be considered one of the most impor-

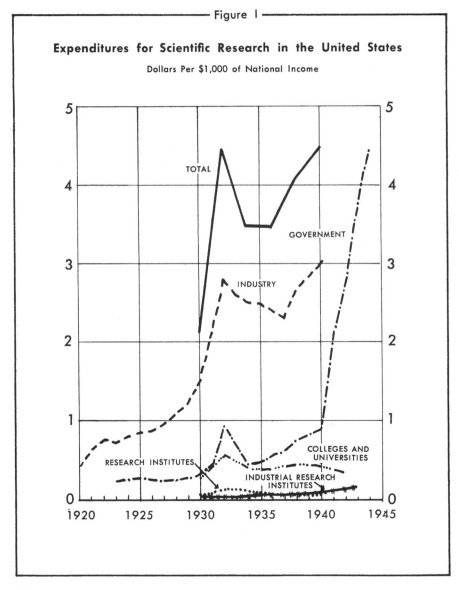

Figure I

Expenditures for Scientific Research in the United States

Dollars Per $1,000 of National Income

tant objectives of a far-sighted national research policy. Idle scientific talent and a retarded rate of scientific and technological progress have been the usual result of economic depression. Steady maintenance of a generally high level of production and employment would naturally obviate the necessity of special stabilization policies in respect to research and technological development. In the period of postwar transition and possible temporary recession, however, increased governmental expenditure may be necessary in order to offset the probable reduction of research activities in industry and in privately financed universities. Even if such a temporary recession should be much milder than the great depression of the early 1930's, the absolute reduction in the national research budget would be substantial since the general level of research is now much higher than it was 15 years ago. It is urged, therefore, that any national scientific foundation that is established should, as far as is consistent with sound and equitable policy, increase its grants for research in periods of depression.

Part Three

Scientific Research in American Universities and Colleges

A. The University as a Research Environment

Historical development has given the sanction of tradition to the prominent role played by the universities in the progress of pure science. The advent of the agricultural and engineering schools has also increased university interest and responsibility in the field of applied research and development.

Several factors combine to emphasize the appropriateness of universities for research. The university as a whole is charged with the responsibility not only of maintaining the knowledge of the past and imparting it to students but of contributing to new knowledge of all kinds. The scientific worker is thus provided with colleagues who, though they may represent widely differing fields, all have an understanding and appreciation of the value of new knowledge.

The long struggle for academic freedom has provided our universities with the means of protecting the scientist from many of the immediate pressures of convention or prejudice. The university at its best provides its workers with a strong sense of group solidarity and security, plus a substantial degree of personal and intellectual freedom. Both are essential in the development of new knowledge, much of which can arouse opposition because of its tendency to challenge current beliefs and practices.

1. *Present Status of Pure Research in American Universities*

The rapid expansion of university education in this country during the present century is encouraging, but it is wise to remember that a large part of this increase has been devoted to undergraduate departments and was especially designed to meet increased teaching responsibilities. More than any other country in the world, the United States has undertaken to provide higher education on a broad base. There has been an even more remarkable percentage rate of growth in the number of students taking postgraduate courses in American universities. Development of research has not, however, paralleled this rapid expansion in teaching.

2. *Backgrounds and Trends of Financial Support to Universities*

The support of our large private universities and colleges has come mainly from endowment gifts and foundation grants. The prodigious growth of our public institutions has been supported by appropriations from the various legislatures. Since 1929 fundamental changes in the American tax and income structure and decline in interest rates have slowed down the rate of new private gifts and endowment earnings. Students' fees constitute an increasing proportion of the total support of private institutions, and capital outlays in those universities are definitely on the downgrade. At the same time research has become increasingly expensive and many State legislatures are finding it difficult to provide adequate support for such activities in their universities.

The science departments of universities have found it necessary, in view of the decrease in gifts by individuals, to rely more upon industrial corporations for assistance. This may imply the distortion of university research in the direction of short-range problems at the expense of more fundamental research. Also the freedom of the university scientist may be decreased by the introduction of some degree of commercial control. Undoubtedly, if proper safeguards are maintained, cooperative research performed for industry in universities can be expected to increase in the future to the advantage of both parties concerned. But in this report which wishes steadily to emphasize the need for freedom in science, it is well to speak of the need to guard against control of science by industry as well as against control of science by government.

If university research is to keep pace with the growth of our economy, if able people are to be attracted to college research and teaching, it is clear that new sources of financial support must be found. Incomes of other professions — doctors, lawyers, dentists, engineers, etc. — have increased considerably during the war. The cost of living has risen markedly. For the first time the personal income tax bears sharply on the middle-income groups. And in the face of these factors, professorial earnings have been frozen at a level which was not considered to be overgenerous 15 years ago. A continuation of this trend will certainly have an adverse effect upon the recruitment and retention of able university scientists.

3. *Immediate Effects of the War*

The above trends were in evidence even before the war. Yet in addition, the present conflict has added a number of very special problems which will dominate the situation for a number of years to come.

First, and most important, has been the virtual cessation of training of new scientific personnel. As a result, we must simply accept as axiomatic the fact that there will be an insufficiency of fully trained young scientists after the war and that it will take a considerable period of time to repair the deficiency.

The war has also created a serious problem of reconversion and rehabilitation for the individual scientist. The mature scholar, as well as the advanced student whose curriculum had to be interrupted, needs re-education and readjustment. The problem is not simply one of returning to the *status quo ante bellum*. In many cases the war has increased the research time and opportunities of American college scientists. Necessarily the concentration in relatively few centers of the bulk of war research will, and should, be reversed in time of peace. But it would definitely not be in the national interest if the dispersal of research staffs away from the largest institutions should mean a marked reduction in the research opportunities and effectiveness of university scientists.

Teaching and research are complementary activities, each aiding and reinforcing the other. But if too much of the teaching is of a routine, elementary character, and if the number of teaching hours is so great as to absorb too much of the time and creative energy of the scientists, then the two activities become competitive. Before the war, in all but a few of the most prosperous universities, teaching loads were excessive from the standpoint of optimal research output.[1] The need to make up for the wartime deferment of training may necessitate the "acceleration" of educational programs to a three-term basis for some years after the war, again with harmful effects to research.

To the well-trained university scientist now engaged in war work, the immediate academic outlook may have lost some of its appeal, especially if he is research minded. For during the war he has had, perhaps for the first time in his life, the facilities and assistance to carry on research in a really efficient way. At the same time, industrial laboratories will be bidding eagerly for his services. University salaries tend to be low compared with those in industry and there has been a steady flow of university scientists into industrial laboratories. Since one of the most important fruits of pure research is the creation of outstanding applied research men, it is very much in the national interest that this movement take place in some degree. But in the immediate postwar period there is danger that an undue number of trained individuals may go into industry, stripping the universities of those who are most competent to teach a new generation of research workers. It is of the utmost importance, therefore, to maintain a favorable competitive position for universities relative to industry.

Paradoxically, increasing the teaching load of university scientists to meet postwar demands may intensify the teaching shortage through its tendency to encourage transfers to industry. A number of partial solutions suggest themselves, each being possible only if financial support is available to make research opportunities more plentiful and teaching more attractive. Numerous scientists on war work may be encouraged to return to their universities; many of the newly trained war scientists may be encouraged to remain in the universities; competent scientists who before the war were in institutions with very little science teaching may be transferred to the more active centers; finally, universities may decide to alter teaching methods and size of classes for at least a few years after the war. These are important short-run makeshifts. In the long run the solution will be found in the training of more scientists.

It is also vitally important that sufficient laboratory assistance, materials, apparatus, clerical and manual aid be provided for those university staff members who are undertaking research in the natural sciences. It is anomalous, to say the least, that universities and colleges should hire first-class scientists, equip them with offices and laboratories, and then fail to provide them with the supplementary funds necessary for productive research. No industrial laboratory would be so imprudent as to use the time of highly paid staff members for doing shop work.

A survey was made by this committee to make possible a quantitative comparison of the support of research in universities, nonprofit research institutions and industrial laboratories during the prewar years. The aim was to discover just how nearly the universities were approximating the

---

[1] See appendix B.

practice, in the use of research personnel, which research institutes and especially industrial laboratories had found to be economical and efficient.[1] The results showed that, just prior to the war, university science departments were spending on the direct operating costs of research—apparatus, materials, technical assistance—sums of the order of 10 to 40 cents for every dollar of salary paid to members of the research staff. A figure of 15 to 20 cents per salary dollar was typical of most departments. Figures above 30 cents were uncommon and were considered to be distinctly liberal by standards of current university practice. By contrast, in a similar survey of a number of industrial research laboratories, a figure of 40 cents for direct research expenditures, per dollar of salary paid to the research staff, was the lowest encountered.[1] In most of the firms reporting, the research expenditures ranged from $1 to $3 for each dollar of salary. Industries have found that generous expenditures for assistance to research workers are economical in the long run. Clearly steps must be taken to help the universities bring their supplementary research expenditures more into line with the best practice.

If the prewar support of research in universities was inadequate, the postwar situation promises to be worse, unless drastic remedies are applied. In view of the wartime increase of some 27 percent in the price level of all goods,[2] each of the already inadequate university research dollars will go much less far than before the war. It is as though every natural science department had already re-

ceived a cut in its postwar research budget of from one-quarter to one-third.

This prospect is serious for all universities but perhaps most acute for the 25 universities just below the first half dozen in size and resources. The large wartime university research laboratories have drawn upon the intermediate universities for staffs. Such tried and experienced men can make valuable contributions in their home institutions if adequate funds are made available. Here lies the immediate opportunity from the Nation's point of view, although much also remains to be done at the highest and lowest levels.

B. Form of Aid to Universities

Although the Government may render important indirect aid to universities by keeping its own laboratories, libraries and other research facilities at a high level of efficiency, the principal need is for direct financial support under conditions which will not endanger academic freedom and the personal independence of the investigator.

This committee does not feel that it is desirable to supply these funds by a series of annual congressional appropriations for specific projects; the difficulties these have raised within the Government service testify to the evils that would be introduced into the university environment by this practice. The preservation of academic freedom requires that funds be allocated in a way that would minimize the possibility of external control and would encourage long-term projects.

Experience in other countries and the example of the successful private foundations suggest that a largely autonomous board with a staff of

[1] See appendix B.
[2] B. L. S. *Cost of Living Index,* all items, March 15, 1940, to January 15, 1945.

men trained in science is the most appropriate agency for carrying on this phase of the Government's responsibility for scientific progress. Recommendations for the formation of a National Research Foundation and additional responsibilities with which it may be charged appear elsewhere in this report. The following paragraphs are devoted to an outline of the committee's views respecting its operation in relation to universities.

To give funds intelligently in support of fundamental research is a difficult task and there is no generally accepted rule of procedure. The private foundations follow a number of different policies and are constantly revising their procedures on the basis of accumulating experience. A Government board would have new and perplexing problems stemming from its status as an arm of the Government, and from the fact that the resources at its command would presumably be large in comparison with those of any single university or private foundation. The committee recommends, therefore, that such a board be in large measure free to formulate its own rules of procedure for allocating funds to universities, as long as these do not transgress certain broad general principles.

The most important of these general principles are as follows:

(1) The funds supplied to the universities should be used for the support of significant research with special emphasis on the universities' position as the chief contributor to pure science.

(2) In making grants the board should assure itself that the university has competent and adequately trained personnel to guide the studies.

(3) Grants to universities or to men working in universities **must** be made in such a way as to **avoid** control of the internal **policy of** the university, so that the university and not the board will have full responsibility for the administration of the grant after it is once made.

(4) A constant effort should be made to improve the general research level in institutions of higher education throughout the country.

No matter on what conditions money is given to universities, the very existence of such support will, of course, modify university policy. In fact, the increased emphasis on research, which will be the object of the Foundation, itself constitutes a change in policy. And despite the fact that our committee is concerned only with the natural sciences, action along the lines proposed cannot fail to have influence on the humanities and the social sciences. It is our **hope** and belief that the provision of funds for the natural sciences would, in some measure, free university funds for use in the other fields. Aside from such general influences, however, it would be necessary to devise ways and means of allocating funds in large measure *without determining what particular problems are to be worked on and who is to carry them out.* The principle of variety and decentralization of control is nowhere more important than in scientific work, where the fostering of novelty must be the first concern. One of the most useful ways of preserving these opportunities is to allow the greatest possible latitude to the accumulated wisdom of university administrative officers and faculties.

The committee has given a great deal of thought to the technical form in which Government grants should be made in order best to reflect these principles. It does not wish to recommend that the proposed board be restricted to the use of any particular plan, as experience will undoubtedly reveal in each defects and advantages which cannot be predicted beforehand. The committee, however, feels that any instrumentality set up to aid research in universities should be empowered to allocate funds in any or all of the following ways:

1. Matching Grants to Private and State Supported Institutions

It is proposed that research funds be made available to accredited universities, colleges, and engineering schools on a matching basis, and in a manner that will be virtually automatic. These grants would be contingent upon satisfying the administrating board that certain clearly stated requirements, largely of a technical bookkeeping nature have been met by the particular institution in question. The grant would be for fluid research purposes within the institution, rather than to any particular department or person. Once a university were accepted as a participant in this plan, and as long as it continued to meet the required conditions, it would expect to receive the grant as a regular annual appropriation, with no other control than some form of Government audit to insure that the money was, in fact, used in support of research. The Government would match dollar for dollar (or according to any other simple formula) the sums the university expended for research.

Although certain practical difficulties must be recognized and dealt with, many considerations make such matching grants attractive in principle. First, and perhaps most important, it leaves to the recipient institution complete freedom in the selection of research programs and personnel. Second, it encourages local support and utilizes the important forces of local interest and pride, both in screening out unworthy projects and in carrying through worth-while ones. Third, the size of the grant is geared more or less realistically to the ability of the institution to utilize it effectively. Fourth, since the grants are largely automatic in character, the board is freed from the burden of investigating intensively the large number of potential recipients and arriving at a decision in regard to the merits and defects of each. The experience of the private foundations demonstrates that judgments of this sort are extremely difficult and time-consuming, even when pursued on a small scale. The burden of work for a Government board with much larger funds at its disposal is bound to be far greater.

There is, of course, the practical problem of determining research costs as distinguished from other outlays. University accounting practice is by no means uniform and there are inherent difficulties in deciding what part of the costs of laboratory space, staff salaries, administrative overhead and so forth is occasioned by research and what part by teaching. Certain funds now received by universities, notably as a result of contracts with industry, should almost certainly not be matched by the Government, especially if the resulting discoveries were to become the exclusive property of the industrial donor. Difficulties of this nature, however, are not insuperable and should not weigh heavily against the many advantages of the scheme.

2. Discretionary Grants

Matching grants, however, may well be attacked as a method of maintaining the status quo, in which a few universities tend to dominate scientific research. It is, in fact, essential to the healthy growth of science that the Foundation should help to spread the research spirit as widely as possible throughout the United States. If the recruitment of future scientific personnel is to proceed from a sufficiently broad base, it is important that as large a number of students as possible be made aware of the research point of view. Many of our colleges and engineering schools are not now able to support a significant amount of research.[3] The level of research practice in these institutions can best be raised through discretionary grants.

The committee recommends, therefore, that the board of trustees be empowered to expend a substantial part of its funds on a discretionary basis, either as grants-in-aid for promising special projects or in the provision of large and expensive capital facilities.

a. Grants-in-aid

Much of the funds now granted by private foundations to universities is in the form of grants-in-aid for special purposes. These range in size from a few hundred dollars for 1 year to several hundred thousand dollars for a period of 5 to 10 years.

A Government foundation with larger sums at its command would presumably be in a position to make appropriations of considerable size and for long periods of time. Indeed the very magnitude of its responsibilities would require that it abstain from frittering away its efforts on a

---

[3] See appendix B.

large number of small and transitory projects.

Once proper precautions are taken to avoid obvious pitfalls, several advantages of relatively stable grants-in-aid argue strongly for their adoption in certain circumstances. Of first importance is the fact that they offer what is probably the best means of supporting promising projects in institutions whose present status does not enable them to benefit sufficiently under matching grants. In this connection particular attention should be given to attaining a better balance of research activity throughout the country.

b. Grants for Capital Facilities

It has already been pointed out that any plans for expansion of research in educational institutions will require additional investment in buildings and equipment. Increases in the total number of students are expected to bring student enrollment considerably above prewar levels and will bring great pressure on existing facilities which are already overcrowded. Substantial sums will therefore be required to provide adequate facilities for advanced research.

In addition the trends in many fields of scientific research point toward the increasing importance of large and highly expensive pieces of equipment which, at present, can be purchased and maintained only by favored institutions. The astronomers were perhaps the first to face this problem, but now the physicist wishes to work with a cyclotron or betatron, and the biochemist with an ultracentrifuge or mass spectrograph; and workers in many fields have need for the services of computing centers or for the use of complicated calculating equipment, such as the differential analyzer. Much of present-day engineering research requires large installations of a semi-industrial nature.

It is, accordingly, suggested that the Government could greatly aid the course of both pure and applied research by making available these facilities to universities, with provisions that they should be used cooperatively by other institutions in the region. A detailed proposal for the management of such facilities will probably need close study; and the needs of each center should be adapted to its peculiar local circumstances.

This proposal appears to have a number of inherent advantages: (1) It provides necessary facilities that would not otherwise be readily available, and an economical and democratic way of using them, (2) it recognizes the cooperative aspects of modern research and provides facilities where workers could come together for a common effort and interchange of ideas, and (3) care in the placement of such equipment would immediately stimulate and strengthen research efforts in hitherto less favored areas.

3. *Postdoctoral Research Fellowships*

Another committee under the chairmanship of Dr. Moe[4] has made a careful analysis of the problem of recruiting and training future research workers up to the level of the doctorate. The Moe Committee is recommending a substantial program

[4] The Moe Committee was appointed by Dr. Bush to assist in answering the President's question "Can an effective program be proposed for discovering and developing scientific talent in American youth so that the continuing future of scientific research in this country may be assured on a level comparable to what has been done during the war?" (See President's letter.)

of undergraduate and predoctoral science fellowships. We should like to reinforce these recommendations by stating our belief that the need for additional personnel is one of the most pressing which faces universities, industry, and Government. The very heart of any successful program of research is the existence of a strong body of highly trained men. Adequate funds can be of immense value in giving a large number of qualified persons the opportunity for the necessary training and study. Not only will provisions for undergraduate and predoctoral fellowships help supply future workers, but grants in the latter category will immediately contribute to the productive research done in universities. Much of the actual experimental work carried on in these institutions is done by students pursuing the Ph.D. degree under the direction of mature investigators. Every additional qualified student assistant thus increases the effectiveness of the senior staff members.

The Bowman Committee also wishes to recommend a program of post-doctoral fellowships as a direct aid to research. The National Research Council, with funds received from the Rockefeller Foundation, has for many years granted a number of fellowships to research workers who have recently received advanced degrees and wish a year or two more to establish themselves firmly in investigative work before taking up extensive teaching responsibilities. A notably high proportion of the recipients have gone on to distinguished careers in science or one of the allied arts, notably in medicine. One of the most important aspects of these fellowships is that their holders have in the majority of instances used them

97

for work at institutions other than those in which they obtained their degrees. Thus, they not only broadened their own training but contributed greatly to the interchange of ideas and methods between laboratories. In the immediate postwar period, an increase in the number of these fellowships would be especially important in re-establishing in scientific work many men who had completed their formal education before joining the armed forces and would thus be ineligible for aid under the G.I. Bill of Rights. The fellowships should also be helpful in certain fields of pure and applied science where a combination of skills is required and where the cost of a thorough training is prohibitive under present conditions.

4. Senior Research Fellowships

Although scholarship and fellowships such as those described above have operated successfully on a fairly large scale in the past, fewer opportunities exist for similar aid to the mature investigator. One of the foundations has for several years given special attention to this field, and shortly before the war the National Research Council instituted the Welch Fellowships in Medicine for men of relatively advanced though hardly mature academic status. In the opinion of the committee, however, much more needs to be done to enable really experienced investigators to develop and utilize their talents most effectively. The problem, in fact, appears to be far beyond the means of private resources. Research workers who have reached the status of assistant professor or above tend to remain in their own universities and their time available for research tends to become increasingly broken up. In theory, the sabbatical year gives an opportunity for intensive research or travel, but in recent years universities have been less and less able to grant such freedom from academic routine. The resulting immobility of the senior staff serves to isolate the intellectual life of a university from that of its fellows, and the individuals concerned, lacking outside stimulation, may incline more and more to perfunctory performance of routine duties. The tendency of American universities to select full professors and department heads from within their own staffs only aggravates these undesirable conditions.

Fellowships large enough to meet the salaries of advanced academic personnel for periods of intensive research work at their own institutions or at other universities would be an effective means of attacking these problems. Such grants offer an especially powerful tool for building up research in institutions that are just beginning to develop the research spirit, either by enabling their faculties to receive advanced training elsewhere or by bringing distinguished workers to them from other institutions. An accompanying grant to cover the use of research facilities should be made to the institution selected by the recipient of the fellowship.

Efforts should also be made to encourage mature scientists in industry and government to avail themselves of the opportunity provided in this program to do fundamental research in universities of their own choice. This should help in part to speed the transition between pure research and its practical applications.

Part Four

Scientific Research in the Government Service

An analysis of the activities of the various scientific bureaus gives convincing proof that the recognized responsibilities of the Government in scientific research are wide indeed. The types of research in which it is directly engaged may be roughly classified under three headings: (1) research that is essential to the effective operation of Government departments; (2) research of broad scientific and economic importance that has long-range value to the Nation and for which the Federal Government has assumed a large share of the responsibility (particularly important has been Government research for industries made up of many small units); and (3) technological research of public concern, which is either too expensive or whose success is too problematical or too far distant to attract the research efforts of commercial enterprise. In this category would also be placed research programs, requiring elaborate coordination, which the Government is peculiarly well-fitted to direct.

Much of Government research is of wide scope and long-range character. It is predominantly a team affair, and often involves the correlation and integration of a Nationwide effort, with the Government enlisting the cooperation of investigators from industry and universities throughout the country. The development of the contract mechanism for sponsoring research has been a most important factor in this type of cooperation. The planning, organization and successful administration of such far-reaching research programs often raise problems much more complex than those encountered in the operation of laboratories devoted exclusively to specific sciences.

The general problem of improving the conditions under which the Government conducts research, and the special problem of coordinating the various scientific activities of the Government, has been previously considered by a number of other committees who have reported to the Congress or to the President. In 1884, a committee of the National Academy of Sciences reported to the Congress on the condition of several of the most important scientific bureaus. This committee recommended the consolidation of the four agencies under consideration into a single Department of Science, or, if that were not deemed practical, the creation of an advisory "permanent commission" charged with coordinating and improving the scientific services of the Government. Neither recommendation was acted upon. In 1908, another committee of the National Academy recommended a permanent board to advise on the work of the scientific bureaus, the board to consist of the heads of the various bureaus, four delegates from Congress, and "five to seven eminent men of science not connected with the Government service." No action was taken as a result of this report. A temporary Science Advisory Board was appointed by President Roosevelt in 1933 and asked to consider specific problems of the organization of various scientific bureaus and to recommend a program for more active support of research by the Federal Government. In the 2 years of its activities, the board made many valuable recommendations and brought about useful improvements in the Government service. The board submitted its first report in 1934 and its final report late in 1935. In this final report the board strongly recommended the creation of a permanent science advisory board for the scientific services of the Federal Government. Several years later the National Resources Committee published a study of Federal aids to research and of the place of scientific work in the Government.[1] Findings of these two committees have been consulted freely in the preparation of the present chapter. Suggestions from research workers and research directors long connected with Government bureaus afford a basis for recommendations supplemental to those proposed by the earlier committees.

A. Suggested Reforms

The special problems of the conduct of research by the Government are made more difficult than is necessary by the application to research activities and to research personnel of regulations designed primarily to govern custodial, regulatory, or other functions of Government. Many of these regulations and restrictions seriously hamper successful prosecution of research work by Government agencies. If research is to be conducted by Government, its distinc-

[1] *Research—A National Resource. I. Relation of the Federal Government to Research.*

tive character should be recognized, and it should be freed from as many as possible of these hampering restrictions. Fiscal and budgetary procedures should be modified to fit the particular needs of research work rather than attempting to adapt research procedures to inflexible regulations applicable to other items of Government expense. Civil Service regulations should be modified to permit the most advantageous procedures for recruiting and classifying scientific personnel. Research by Government bureaus should be coordinated with research in other public and private scientific institutions.

1. *Fiscal and Budgetary Procedures*

The scientific work of Government bureaus could be assisted greatly by simplifying procedures in order to permit more effective use of the funds appropriated for research. The principal modifications suggested here are aimed at granting wider latitude and greater flexibility for planning and executing sustained research programs. The necessary changes in procedure can probably best be determined by a special committee composed of governmental and nongovernmental scientists and representatives of the budgetary or appropriating authorities.

a. *Appropriations for Long-Term Programs*

Current budgetary procedure of Government provides funds on an annual basis, yet only a small percentage of the research conducted by Government agencies can be planned adequately or appraised satisfactorily on a 1-year basis. Research programs should somehow be assured in terms of their long-run objectives. If approved, funds should be guaranteed over the period of years necessary to permit continuity of effort and attainment of these ultimate objectives. Appropriations should be in lump sums for broad programs rather than in specific sums for detailed projects. Requiring detailed justifications of an annual budget tends to stultify research by ignoring its intrinsic uncertainty. Appropriations within the assured sum might then be made available as at present in the annual budget. This plan has attained limited acceptance in certain departments of the Government, but the procedure should be made uniform throughout scientific bureaus. It gives needed flexibility to research programs and permits modification to meet unexpected developments which almost inevitably arise.

b. *Cooperative Support of Research by Public and Private Agencies*

The degree of cooperation by public and private agencies in the financial support of research has never been uniform in all departments of the Government. It should be made a relatively simple matter for any scientific bureau of the Government to accept funds from State or local governments, from nonprofit research institutions, or from private industry, for cooperative scientific investigations that are in the public interest.

It is particularly important that Federal research agencies should be able to cooperate freely with State and municipal governments. Many problems of predominantly local concern can be studied most advantageously by State agencies, such as agricultural experiment stations, health departments, and mineral resource bureaus. To the extent that the results of these studies are of more than local interest, they should receive financial support through the Federal bureaus that are particularly interested. The Federal Government should, perhaps, make a special effort to stimulate development of research organizations in backward States.

c. *Simplification of Fiscal Regulations*

Government regulations regarding the purchase of supplies and equipment, while intended to assure economy and fair dealing to all, often hamper research programs. The calling for bids and the insistence on purchase of the lowest-priced material is no doubt fully justified for the great bulk of Government supplies. Nevertheless, the required procedures do not always yield scientific equipment of the best quality, and the nominal saving is usually far outweighed by intangible losses in delay and frustration of the research staff. Liberalization of the rules for purchase of scientific equipment is, therefore, recommended.

2. *Operation of the Civil Service*

The most important single factor in scientific and technical work is the quality of personnel employed. Scientific and professional personnel in Government service are now subject to approximately the same system of recruitment, promotion, and supervision as those in the clerical, fiscal, and custodial positions. Separate and distinct procedures for recruiting and classifying scientific personnel are warranted by the exacting technical requirements in these services. No one change from current practice would do more to improve the quality of research conducted by the Government than the establishment of a separate branch of the Civil Service for scientific and technical positions.

The Civil Service was instituted to replace the demoralizing "Spoils System" by an orderly merit system of recruiting efficient personnel for Government service. It has been largely successful in eliminating the "spoilsmen"; and any modifications designed to improve present methods of recruiting and protecting personnel must not imperil the defense now afforded against political influence and favoritism in making appointments. The keystone of the merit system is competition open to all qualified applicants for a position; but the Civil Service has been severely criticized because of the slow and cumbersome machinery necessary to insure this competition. The general suggestions offered here are designed to meet the more serious of these criticisms without undermining the essentials of the merit system. The precise form of changes needed in present Civil Service procedures deserves study by a special committee of governmental, university, and industrial scientists and representatives of the Civil Service Commission.

a. *Entrance Requirements for Scientific Service*

The standards for entrance into scientific and professional positions in the Government should be approximately those maintained for comparable posts in universities and industries. Civil Service positions are subjected to continuous political pressure to relax entrance requirements; and recently the educational requirements for a number of scientific classifications have been removed. This opens the way to possible appointments by personal favoritism and political preference. Action should be taken immediately to re-establish the requirement of a university or college degree for entrance into all scientific and professional services. Exceptions in especially meritorious cases should be granted only upon recommendation of qualified scientists.

In many types of Government employment, standards are not lowered by granting military preference to candidates who have served in the armed forces, although, strictly speaking, such preference is a departure from the merit system. In scientific and technical services, however, individuals unable to qualify without special preference are not really benefited by appointment to positions for which they are unqualified; and when such appointments are made, the work inevitably suffers. Any lowering of entrance requirements, whether for civilians or veterans, is distinctly harmful to the scientific services of Government.

b. *Recruitment of Scientific Personnel*

The methods of recruiting for governmental service presupposes a supply of able applicants for every position to be filled. However, in the years immediately preceding the present war there was a shortage of able young scientists. This shortage is likely to be even more acute after the war, because of the interruption of training programs.

Government scientific bureaus are under a severe handicap in competing with industrial laboratories which employ college seniors by the use of the personal interview followed by prompt appointment. This handicap should be offset as much as possible without jeopardizing the fundamental objectives of Civil Service. It should be permissible for representatives of Government agencies to interview students and to persuade the more able ones to apply for employment. The months of delay between application, examination, and notification of appointment should, if possible, be reduced to a few days. College seniors could be given appointments effective on the day of their graduation. The necessary safeguards could be maintained by a longer period of probationary employment and by the requirement of suitable examinations before promotion to higher grades. Government bureaus could further improve their chances of successful recruitment from the colleges by making wider use of temporary student appointments during summer vacations.

c. *Salary Scale*

The opportunity for full-time scientific work, freedom to publish results, and the satisfaction of serving the national interest attract many able scientists to the Government service. But salary scales must be broadly commensurate with those of private institutions if these scientists are expected to remain in Government service. Entrance salaries in Government scientific positions are usually slightly above this competitive level. However, promotion is slow in Government service, and the higher positions carry salaries much lower than those offered in industry. The present system of efficiency ratings and promotion procedures is designed to assure fair and uniform treatment for all governmental positions. But this system is so elaborate that it requires handling by many persons of nontechnical training. Furthermore, in most branches of Government service, the higher salaries are almost solely for supervisory positions. As a result, the senior professional position, with a salary range of $4,600 to $5,400 a year, is the highest ordinarily attained by Government scientists in nonadministrative positions.

Civil Service regulations should be modified to permit exceptionally qualified scientists to reach salaries of $9,000 or more a year even though they may not have important administrative responsibilities.

It is sometimes said that one of the most serious limitations of scientific work by the Government is the inability to pay salaries large enough to get outstanding directors for research organizations. Although no legal restrictions, except the necessity for congressional approval, prevent the employment of a director of a research bureau at whatever salary is deemed necessary, practical considerations—such as comparison with the salaries of Department Secretaries and members of Congress—inevitably impose limitations. In actual practice, few research directors have received more than $9,000 a year.

Under the stress of war needs, when expert talent was urgently required, the departments of Government, particularly the so-called "war agencies," succeeded to some extent in breaking down the tradition of low Government salaries. Many technical positions and salaries were, in effect, up-graded; and this contributed significantly to the flow of talent into Government service during the war emergency. A more general and permanent up-grading of positions and salaries in the scientific services of Government, accompanied by a careful selection of new talent, would be a major contribution to improvement of the quality of research conducted by the Government.

d. *Security of Tenure*

The security of tenure in Civil Service is partial compensation for the lower salaries in many types of governmental employment, especially during periods of depression. But if scientific and professional personnel are to be classified separately from other Government employees, and if they are to receive salaries approximating those of their colleagues in universities and in industry, care must be taken that this security of tenure does not become a shelter for incompetence and mediocrity.

Many of the more able and energetic scientists in Government service are offered higher salaries elsewhere. Inevitably, a number of these offers are accepted, with the result that the less qualified employees tend to constitute a larger proportion of those who remain. Unless a research bureau can replace such losses with new employees of equal ability, it is forced to operate with only the residue of its scientific staff after continuous raids. An additional handicap is the difficulty, under Civil Service regulations, of demoting or dismissing incompetent, mediocre, or poorly adjusted individuals.

Higher standards for entrance into scientific positions, longer and more closely supervised periods of probation, examinations for promotions in the lowest grades, with the alternative of separation from the service, and higher salaries for the abler scientists are some of the methods by which the quality of scientific work of the Government can be improved.

3. Coordination of Governmental Research

The extensive development of the sciences in recent years, and the increasing complexity of governmental research, make it more difficult each year to coordinate the scientific work conducted by the Government and to integrate governmental research with that of universities, endowed institutions, and industrial organiza-tions. Parallel investigations of certain important research problems are to be encouraged rather than avoided, and duplication should not necessarily be the bugbear in science that it is in other types of governmental activity. Nevertheless, it becomes increasingly important that the research personnel of various governmental bureaus keep in close touch with one another and with current technical developments and public needs.

a. Coordination of Research Within the Government

A specific need is for an inter-bureau committee or council of representatives of the principal scientific bureaus. Such a committee might be set up under the Bureau of the Budget, or other appropriate auspices, to advise on interrelationships of research programs of the different agencies, and to compare the effectiveness of different procedures for administering governmental research. Recommendations from such a committee on policies of budget procedure or of classification of scientific personnel should carry more weight than the recommendation of a single bureau.

The practice of utilizing scientific employees of one bureau as consultants for other bureaus is difficult under existing regulations. But if this practice were generally adopted, it would further coordination of research programs by disseminating more widely a knowledge of the related problems under investigation by various agencies and of the different methods by which these problems are being attacked.

b. Coordination of Governmental Research with Outside Organizations

There is a widespread impression that a research project, once started by a Government bureau, may continue long after it has served its original purpose. Research projects need continuous reappraisal in the light of scientific advance and technological developments. Orderly revision of research programs should be the normal and expected result of scientific progress. The danger that a research bureau may fail to revise its programs or its methods when they become obsolete is minimized most surely by encouraging members of the scientific staff to maintain close contact with their professional colleagues elsewhere. Government employees engaged in research should be encouraged to participate in the activities and publications of national scientific societies. This means, among other things, more liberal funds for travel to scientific meetings. Furthermore, it should be legally possible for any Government bureau to keep in close touch with modern ideas within its field of science by assigning employees on full pay for graduate work at universities or for research projects to be conducted at endowed or industrial institutions or at official research organizations in this or other countries. Scientists from universities, research institutions, State agencies and industrial organizations should be invited to accept appointments for short-term projects in Government bureaus. Facilities should likewise be extended to visiting scientists from foreign countries.

c. *Clarification of Patent Policy for Government Employees*

The present policy of granting patents to the employees of some Government bureaus for inventions in the field of the bureau's official duties does not instill public confidence in Government employees nor encourage industry to share new information with Government agencies. Attention should be given to the recommendations of the National Patent Planning Commission that all inventions made within the specifically designated duties of Government employees be assigned to the Government and that doubtful cases be decided by a central board on Government patents.[1]

4. *Advisory Committees to the Separate Bureaus and a Permanent Science Advisory Board*

Many of the changes here recommended to assure proper coordination of governmental research and raise the level of its performance depend in considerable measure upon the existence of advisory committees to the several scientific bureaus. The excellent service rendered by the several advisory committees already in existence has demonstrated the value of these bodies. Their use, however, is not universal, and at present only the more progressive bureaus actively seek outside advice. It is therefore urged that advisory committees, composed of scientists from outside the Government service, be established for each of the bureaus or agencies in which extensive research is being conducted.

There has been ample experience, also, to demonstrate the need for a permanent Science Advisory Board, similar to the body which served so successfully on a temporary basis from 1933 to 1935. Such a central board could correlate the activities of the specialized advisory committees, and would probably be the proper

[1] See *Second Report of the National Patent Planning Commission* (Washington, 1944), pp. 10-12.

body to recommend the personnel of the various committees. It would be in a position to advise Congress and the Bureau of the Budget on the quality and importance of research being conducted by the bureaus. By being able to rely upon the disinterested advice of such a body, Congress might be willing to appropriate sums for long-term programs of basic research whose importance it is difficult or impossible for nontechnical persons to evaluate properly. Besides consulting with the bureau chiefs on their individual or collective problems, the board would find it helpful to meet at stated intervals with the interbureau committee proposed earlier in this chapter.

We add our recommendation to those of earlier committees and strongly urge the creation of a permanent Science Advisory Board, empowered to assume over-all responsibility for advising the various branches of the Government in scientific matters. We suggest that this board cooperate closely with the National Research Foundation.

Part Five

Aids to Industrial Research and Technology

Industrial research in America has enjoyed a rapid and extensive growth. There are also widespread indications that industry is planning to undertake applied research on a greatly expanded scale in the postwar period—an encouraging and wholesome prospect. At the same time it is evident that research in American industry is concentrated to a considerable extent in a relatively small number of industrial units and in a few particularly progressive industries. Thirteen companies employed nearly one-third of all industrial research personnel in the year 1938. In the rubber industry, one-quarter of the companies employed 90 percent of the research workers, while in petroleum and industrial chemicals the respective percentages were 85 and 88. This is not to suggest that there should be a considerable degree of uniformity among the units of an industry or between industries as to the percentage of research effort in each. But the implications of the increasing concentration of industrial research in this country deserve special study.

One important fact is clear—the process of transition from pure research to its practical application does not work equally effectively in all industries. For example, the petroleum industry has for years supported far more research than has the coal industry. New technical developments in the petroleum field have made it possible to carry on an increasing amount of research while the relative backwardness of the coal industry, where small units predominate, has resulted in fewer and fewer new developments and a less and less healthy over-all situation.

Time did not permit an intensive and well-rounded investigation of this subject. The Committee feels strongly, however, that the National Research Foundation should be charged with the responsibility of studying the process of technological development in industry and of experimenting with methods of aid to industrial research. The following suggestions are tentative and submitted with the thought that they might be of assistance to the Board in meeting this important challenge.

A. Assistance to Technical Clinics for Small Business Enterprise

It is the belief of the Committee that the most effective research wells up from below. Our objective, therefore, should be to develop as many individual centers of research initiative in industry as is possible. The seeds of industrial research that are planted now in small, vigorous industrial enterprises may yield tremendous returns in the future. There is considerable difficulty, however, in getting research started in enterprises which have not been research-minded in the past.

To meet this need a number of special research clinics have been established in different regions, *e. g.*,

107

the New England Industrial Research Foundation. These clinics make their services available to the small business concerns of the region in which they serve. It is difficult to place this type of enterprise entirely on a self-supporting basis especially where its important promotional activities are concerned. The Committee believes that this movement should be encouraged. It therefore recommends that the National Research Foundation be empowered to make sustaining grants to cover part of the administrative costs involved in such organizations, provided they are run on a nonprofit basis.

The activities of such services should include stimulation of business interest in research and technical developments, aid to small businesses in interpreting the trends in technical developments, consultation with individual concerns to aid them in a diagnosis of their technical problems, and maintenance of a directory service to put small businessmen in touch with competent individuals and proper sources of information necessary for further work.

Universities, engineering schools and nonprofit industrial research institutes should be eligible to receive grants from the National Research Foundation to perform such services. Insofar as possible, organizations with grass-roots foundations standing high in community prestige and offering a substantial background in active research work, should be selected. The staff of such a clinic might include a limited number of full- or part-time individuals for general promotional and advisory work; but there should also be available a panel of experts in as many fields as possible for counsel on specific problems.

War experience has demonstrated that such organizations must be able to bring their information directly to the plant. It is believed that in the proper hands such services will prove very helpful and can be of significant value in the long run in developing vigorous new research organizations and reducing the existing concentration of research in a relatively small number of companies.

B. Grants to Nonprofit Industrial Institutes for Fundamental Research

In recent years an increasing number of industries in which research has lagged in the past have attempted to meet the problem by establishing special research institutes to serve the industry. Such institutes are usually supported by annual grants from individual business concerns. One of the difficulties that these institutes have faced is the pressure for short-range accomplishments. In consequence, research undertaken has not been sufficiently basic to achieve the most significant results. It is suggested, therefore, that where nonprofit industrial institutes are deemed capable of undertaking important long-range research they should be eligible for grants for fundamental research from the National Research Foundation.

C. Encouragement for New Scientific Enterprises

In addition to these recommendations, some members of the Committee feel that special steps should be taken to encourage the launching of small scientific enterprises. Other members, while sympathetic to these objectives, do not believe that any practical method could be devised for handling such problems through a Government agency.

108

Those members of the Committee who favor taking some positive action to help launch new scientific enterprises believe that greater opportunities should be provided to individuals who are primarily interested in new applications of recent advances in pure science rather than in basic inquiry itself. This thought has been elaborated by one of the members of the Committee in the following terms:

The country needs new types of industrial activity. We should not be satisfied with the cycle of displacement of one good technical product made of metal by the same product made of plastic, and so on, in a rather unimaginative utilization of fundamental developments. What is required is the rapid invention and evolution of the peacetime analogues of jet-propelled vehicles, bazookas, and the multiplicity of secret, bold developments of the war.

New types of industrial activity could be aided if students of engineering and science were strongly encouraged at the undergraduate stage to study unsolved technical problems and to invent solutions for them. On graduation those young men who wish to strike out for themselves should have the opportunity to complete their inventions, both theoretically and practically, in an actual enterprise. In large industrial organizations which provide the principal outlet for such men there is a long path of duty which the young scientist must pursue before he can become very effective in original contribution. Furthermore, most large industrial concerns are limited by policy to special directions of expansion within the well-established field of activity of the company. On the other hand, most small companies do not have the resources or the facilities to support "scientific prospecting." Thus the young man leaving the university with a proposal for a new kind of industrial activity is frequently not able to find a matrix for the development of his ideas in any established industrial organization.

Neither is it always satisfactory that such a potential scientific entrepreneur remain in the university for graduate work. The Ph.D. degree in the American university may not best fit a man for such a career; it makes him a good scholar but may dampen his early leanings in the direction of the commercial development of his ideas.

The Committee was not able to agree on a solution to this problem. The matter was regarded as of sufficient importance, however, to justify careful investigation by the National Research Foundation in the hope that it might be able to devise special methods and techniques of encouraging young scientists in the development of their inventions and in the launching of new scientific enterprises.

D. Strengthening the Patent System

Patents are the life of research. No study of the aids to research or the incentives to research would be complete without an inquiry into the manner in which the patent laws and the patent system of this country might be strengthened. The Committee has given its attention to this important problem and has advised Dr. Bush informally of its views on this subject.

No detailed recommendations on the patent aspects of research are herein contained since Dr. Bush is independently making a study of this problem looking to a separate report to the President. This Committee wishes to emphasize, however, the very vital importance of a strong patent system to the development of new and active small enterprises and the stimulation of healthy scientific research.

Part Six

Taxation and Research

Federal corporate income taxes have an important bearing on the amount of scientific research and new-product development undertaken by private enterprises. An examination of the present treatment of research and development expenditures for tax purposes is therefore an important aspect of a study designed to determine, as requested by President Roosevelt, what the Government can do to aid research activities conducted by private organizations.

A. Present Tax Treatment of Research and Development Expenditures

The *deduction of research and development expenditures as current charges against net income* is generally permitted by the Bureau of Internal Revenue. In broad terms the policy of the Bureau appears to be as follows: Firms that spend approximately the same amount on research and development work year after year and consistently claim these expenditures as deductions from current income seldom have substantial amounts of their claims disallowed. On the other hand, where the amounts spent on research and development fluctuate widely from year to year and where the taxpayer does not follow a consistent accounting practice in handling research and development expenditures, the Bureau tends to question more closely the taxpayer's treatment of such expenditures. This policy may sometimes result in a less favorable treatment for new and small companies than for large, established companies.

1. *Uncertainty in Minds of Taxpayers*

Many taxpayers believe that in recent years the Bureau of Internal Revenue has been adopting an increasingly critical attitude toward the deductibility of research and development costs. This impression, so far as the Committee can determine, is not the result of any deliberate change in the policy of the Bureau. Rather, it has probably arisen from the justifiable tendency of the Bureau to review more closely all items affecting taxable income in years of high tax rates.

The uncertainty on the part of taxpayers is heightened by the fact that the tax law and Treasury regulations do not clearly specify the proper treatment of research and development costs. There are relatively few court cases on the problem; moreover, existing cases seem to support the view that many research and development costs are capital expenditures. Consequently, if the Bureau should abandon its present liberal policy and attempt to enforce the capitalization of research and development costs wherever possible, its action might well be sustained in court.

If research and development costs were required to be capitalized, they could presumably be amortized over their useful life. The task of determining the proper basis of amortization would, however, be extremely difficult and frequently impossible of solution. If the tax law were very narrowly interpreted, it is conceivable, although not likely, that the difficulty of determining a proper basis for amortizing capitalized research costs might make it impossible for such costs ever to be deducted for tax purposes. The Treasury regulations now permit intangible assets to be amortized only when the useful life of the asset can be determined with reasonable accuracy.

Since a delay of several years ordinarily occurs before a tax return is finally audited and closed by the Bureau of Internal Revenue, the uncertain status of the deductibility of research and development expenditures can involve very large sums of money. Small firms making heavy research expenditures, in particular, are restricted by this uncertainty in their commitments for fixed investments.

2. *Proper Accounting Treatment of Research and Development Costs*

No simple, universally applicable principles can be laid down as to the proper accounting treatment of research and development costs. Some research costs are clearly current expenses: they either turn out to be worthless or merely enable the taxpayer to keep abreast of his competitor. Other research expenditures may improve the long-run position of an enterprise, but the amount of the expenditures properly allocable to a given product, and the proper basis of amortization of these expenditures, may be almost impossible to determine. In a few cases such as, perhaps, the development of a new

model of an airplane, the capital nature of the expenditure may be fairly obvious, and it may be possible to determine a reasonably satisfactory basis for amortizing the expenditure. Even in such instances, however, it is frequently impossible to determine in the year that a given expenditure is made whether a valuable capital asset will be developed.

3. *The Public Interest*

This report assumes that the stimulation of research and development work, especially by small enterprises, is in the national interest. It therefore follows that the present tax uncertainties of research expenditures should be removed. The case for taking this action is particularly strong since, for the most part, all that is needed is a specific legal sanction of the present Treasury practice.

B. Recommendations for Legislative Action

Recommendation (1).—Deductibility of expenditures on research and development (other than expenditures for the acquisition of tangible capital assets). The Internal Revenue Code should be amended to give the taxpayer in every taxable year an option:

(a) To deduct currently all expenditures on scientific research and the development of new products and processes, other than expenditures for the acquisition of tangible capital assets; or

(b) To capitalize such expenditures as deferred charges and amortize them according to a specified plan that in the judgment of the taxpayer is deemed reasonable; or

(c) To deduct currently such part of these expenditures as in the judgment of the taxpayer is deemed to be a current cost and to capitalize the remainder as deferred charges and amortize them according to a specified plan that in the judgment of the taxpayer is deemed reasonable.

This recommendation appears to be clearly desirable. Its primary effect would be to give clear legal sanction to present practice and hence to remove the uncertainty of the present law. A secondary effect would be to give the taxpayer more flexibility in the deduction of research and development costs. Very little change in current practice, however, would ordinarily result from the increased degree of flexibility. Most taxpayers would continue their present policy of treating research and development costs as an annual expense. Since no simple rule, properly applicable to all cases can be devised, it seems wise to give the taxpayer considerable freedom of action.

Recommendation (2). — Amortization of expenditures for the acquisition of tangible capital assets used for scientific research and the development of new products and processes. The Internal Revenue Code should be amended to give the taxpayer an option:

(*a*) To amortize the cost of tangible capital assets used for scientific research and the development of new products and processes in equal amounts over a period of five years; or

(*b*) To depreciate such assets at the same rates as ordinarily allowed on such assets.

Recommendation (2) provides for an optional accelerated amortization of tangible capital assets acquired and used for the development of new products and processes. As such, it is consistent in purpose with recommendations made by the Roosevelt Administration for the accelerated amortization of all depreciable assets. The present recommendation would not, however, raise many of the difficulties of the general proposal, since it would apply to only a very small percentage of fixed assets and hence would not significantly affect the revenue yield of the tax structure.

In general, expenditures for the acquisition of tangible capital assets constitute a minor fraction of all outlays on research and development. Moreover, in some instances at least, research equipment is already depreciated at a rapid rate. Nevertheless, this recommendation has been supported by most of the businessmen whose opinion on the proposal has been obtained.

C. Broad Tax Considerations

The preceding sections of this report have been confined to issues related directly to the treatment of research and development expenditures. Two broader tax revisions, clearly desirable on other grounds than for the sole purpose of promoting research and development work, would be very helpful in stimulating increased research and development expenditures. Consequently, the Committee concurs in the recommendations that have already been made by other groups:

(1) That immediate legislative action be taken to make the tax refunds from the carry-back provisions and from the postwar refund of 10 percent of excess profits taxes more promptly available to taxpayers during the transition period; and

(2) That the net operating loss carry-over provided by the present tax law be increased from 2 years, to, at least, 5 or 6 years.

Part Seven

International Scientific Cooperation

Perhaps more than any other national activity, scientific research and development depend upon close relationships with other countries. Scientific knowledge is not limited by geographical or racial boundaries, and it is almost impossible to think of any branch of science which has progressed very far without amalgamating discoveries made in several different nations. In the past, most of this interchange has gone on informally and directly between the members of the scientific communities concerned, without regard to political considerations. Certain obvious barriers such as that of language have hampered free communication, but, on the whole, relations between scientists have probably been closer than between the representatives of any other segment of society.

The growth of science in the last few decades and its increasingly close relationship to other national interests have demonstrated the need for more official methods of carrying on international scientific activity.

A. Support and Sponsorship of International Cooperative Scientific Enterprises

That this country has never provided any method of participating officially in international scientific enterprises has frequently been an embarrassment to various scientific groups. If the present tendency, in other countries, of closely integrating science with Government continues, the need will exist for some official body to carry on international scientific activities. (A good example of such an activity was the so-called "International Polar Year" in which several countries, interested in the compilation of scientific data of the Arctic regions, pooled information and techniques. This cooperation gave added value to the results of the study by providing uniform methods of observation and presentation of the data collected.) The Government could not only provide some modest financing for such international cooperative projects, but it could also facilitate them by arranging for means of travel, visas, and so forth.

It is therefore recommended that the National Research Foundation be charged with the responsibility of participating in such international cooperative scientific enterprises as it deems desirable.

1. *International Scientific Congresses*

The Foundation should also participate in arrangements for international scientific conferences whereby scientific workers in different countries may be brought together to exchange ideas. These were held more or less regularly before the war and were found to be stimulating and useful forms of assistance to the advancement of science. As far as the United States was concerned, however, its representatives were frequently hampered, especially when they were acting as hosts for conferences held in this country, by the lack of Government financial aid and by difficulties in arranging for official courtesies relating to the travel of outstanding men from abroad and for other marks of official recognition which are commonly available in Europe. An organization such as the National Research Foundation could be very helpful in making these arrangements.

2. International Fellowships

If the Foundation is set up as suggested, it would be empowered to grant scholarships and fellowships to qualified scientists. It is recommended that attention be given to awarding some of these fellowships to Americans who wish foreign travel and study, and to scientists from abroad to undertake advanced research in this country. Private foundations have found this to be an excellent way not only of aiding scientific progress but of promoting international understanding as well. Holders of such fellowships are likely to be disinterested representatives of their countries and well equipped to observe the national life of the country they visit. Such a program should be undertaken with particular care to avoid specialized political or personal interests, and it would be well to draw freely upon the experience of the National Research Council or the private foundations which have been successful in this field in the past.

3. Scientific Attachés

The Committee would like to suggest, as an experiment, that scientific attachés be appointed to serve in certain selected United States embassies. They should be men of high professional scientific attainments whose tenure of the post would be temporary—perhaps 1 or 2 years—and whose principal duties would be concerned with facilitating the various aspects of scientific cooperation discussed above. It is also expected that in less formal ways they would improve their knowledge of science as it is pursued abroad and would in turn contribute something from their experience in this country. Such a post would appear to be most important in countries such as Russia where a great deal, if not all, of the scientific activity is controlled or directed by the government and where other channels of scientific communication have been greatly restricted for several years.

Part Eight

A National Research Foundation

As a means to implement the recommendations of this report, it is proposed that the Congress should create a National Research Foundation. The function of this new Federal agency should be to assist and encourage research in the public interest by disbursement of funds allocated by the Congress for that purpose. Its board of trustees should be eminent men who are cognizant of the needs of science, and experienced in administration. The members of this board should be appointed by the President of the United States from a panel nominated by the National Academy of Sciences.

It is proposed that the Foundation be given an original nonearning capitalization of $500,000,000, to be called and expended, with the approval of Congress, over not less than 10 years. As has been discussed earlier in this report, scientists and educators emphasize the cardinal importance of creating a board which (1) can budget its expenditures over a considerable period of time; (2) will not be subject to review in detail by the legislature, and (3) will be able to withstand political pressures. The British University Grants Committee, which has been operating successfully for many years with funds supplied by Parliament and whose postwar role is undergoing great expansion, serves as one of many examples corroborating the desirability of the above features. It should be clearly understood, however, that the present committee does not recommend any instrumentality which would not be accountable to the President, the Congress, and the public for its operations.

A. Organization

The board of trustees should comprise about 15 members, each serving on a part-time basis, with remuneration at the rate of $50 per diem when on official duty, plus necessary travel and subsistence expenses. The term of office should normally be for 5 years and no retiring member should be eligible for reappointment until after a lapse of 2 additional years. In order to insure continuity, there should be a staggering of the terms of office of the various members. It is suggested, therefore, that at the outset the first appointments be for varying periods of time. Because the progress of science depends in great measure on the vigorous and progressive abilities of younger men, the Committee suggests that in making appointments to the board and in its policies on retirements an effort be made to keep the age distribution such as to assure dynamic leadership.

The board should have the power to appoint an executive director of recognized ability who would be a full-time officer, receiving a salary commensurate with the dignity and importance of his position and responsibilities. The details of the internal administrative organization of the Foundation cannot profitably be discussed here. The executive staff of the Foundation and its board of directors would no doubt wish to modify and change administrative procedures in the light of experience acquired. Presumably, the executive director would have at his command a staff of experts, each responsible for a major division of science: the physical sciences, the biological sciences, engineering, agriculture, etc. These experts would be full-time professional employees of high salary and status, charged with the task of keeping in touch with research all over the country. By lightening the burden of administration for the board, they would permit it to concentrate on important policy decisions. Not the least of the problems which would have to be faced by the new organization would be that of maintaining close contact with research in all parts of the country.

B. Powers and Responsibilities

The broad, general object of the Foundation is to promote the general welfare through support to science. However, more concrete powers, responsibilities, and limitations must, in the judgment of this Committee, be placed upon the scientific Foundation when it is established. Under the guise of "promoting the general welfare," the agency should not be able to set itself up in business to produce in competition with existing industry. Its primary purpose is to provide encouragement, and where necessary, financial aid, without at the same time introducing centralized control of research. The Foundation should be empowered to receive gifts or bequests for the support of scientific research from outside sources as well as from the Government.

In carrying out its objectives, the Foundation should take all necessary or proper steps:

(a) To study and keep itself currently informed on the present state of science in the United States and to seek ways of applying its resources to the discovery of useful knowledge.

(b) To initiate, encourage, sponsor and finance scientific research and development with emphasis on research aimed at (1) increasing the general fund of basic scientific knowledge and thus creating new industries and increased employment, (2) promoting the conservation and better utilization of natural resources, and (3) improving the health of the Nation. The Foundation should utilize to the greatest extent feasible the existing facilities in the Federal Government, State governments, educational institutions, public and private foundations, laboratories, and research institutes. No contract or grant-in-aid made in furtherance of this provision should introduce control by the Foundation over the internal policies or operations of the contractor or grantee.

(c) To establish or provide new or additional scientific and technical research facilities in geographical areas or specialized fields of study or endeavor where none exist or where existing facilities are deemed by the Foundation to be inadequate: *Provided,* That the Foundation should not itself undertake directly to operate such facilities.

(d) To provide for and assure the most comprehensive collection and dissemination of scientific and technical knowledge and information by aids to libraries, bibliographic services, translating activities, etc.

(e) To seek out latent scientific talent, and to foster and support scientific and technical education and training through grants to individuals, educational institutions, public

116

and private foundations, laboratories and research institutes, and through scholarships, fellowships and prizes.

(*f*) To assist scientists, inventors, and research workers by affording them opportunities to engage in research and developmental activities.

(*g*) To act in cooperation with the National Academy of Sciences in advising the President, the Congress, and the various departments, independent establishments, and agencies of the Government on scientific matters.

(*h*) To make its facilities, personnel and resources fully available to the United States of America in the event of war or the declaration of a national emergency by the President or the Congress.

(*i*) To cooperate with the Army, Navy, and civilian military research organizations for the rapid interchange of information on basic scientific problems of use in national defense. It should coordinate its activities wherever possible with these organizations to prevent unnecessary duplication.

(*j*) To assist industry and business, particularly small enterprises, in establishing research facilities and in obtaining scientific and technical information and guidance, in order to expedite the transition from scientific discovery to technological application.

(*k*) To help maintain a continuous and steady expansion of scientific research by increasing its grants for research in periods of depression, as far as is consistent with sound and equitable policy.

(*l*) To represent the United States of America in effecting better international cooperation in scientific activities, to assist in the freer international exchange of scientific and technical knowledge and information and scientific and technical personnel, to help sponsor and finance international scientific congresses or associations and cooperative scientific research programs.

The board should also be empowered to make grants for such historical and economic studies as it believes necessary to fulfill its responsibilities in investigating scientific research and its practical applications in industry.

Finally, it should be stressed that confidence must be reposed in the integrity, character, and qualifications of the individuals comprising the board of directors. No curbs, restrictions, or limitations on their powers would provide adequate safeguards, or take the place of character and ability; and the introduction of a series of hampering limitations would lead to inflexibility and inefficient operation.

C. Patent Policies of the Foundation

The Foundation should set up its own general rules for the handling of patent policies. It is felt that in establishing these policies the Foundation should interfere as little as possible with the practices of the different universities and research institutions.

It is expected that the obtaining of patents by universities on work financed by the National Research Foundation will remain a minor by-product of the fundamental research undertaken. The patent policy of the universities and research institutions should not be permitted to interfere with early publication of results. Moreover, the patent policy of the recipients of grants should be such as to foster widespread public use of worth-while developments.

117

Library Aids

Adequate technical libraries are an indispensable tool for research workers. Every new discovery depends upon a host of former ones, and every year brings additions to the store of knowledge which must be mastered by the scientist. The magnitude of the task of keeping all this knowledge available to the scholar requires that he be provided with every possible aid and convenience. These services may be considered under three headings: (A) interlibrary cooperation, (B) abstracting and translating services, and (C) bibliographic and reference services.

At the present time none of these services can be said to be entirely adequate and the rapid expansion of published materials makes it very unlikely that private resources can continue library services even on their present level. The problem is so large and requires so much special knowledge and training that the Committee does not feel in a position to make specific recommendations as to where Government aid can best be brought to bear. It does, however, feel strongly that a Government board such as the National Research Foundation proposed elsewhere in this report should give careful attention to the problems presented below, and should devote part of its funds to their solution.

Several existing Government agencies, such as the Library of Congress and the Army Medical Library, could, if they were supplied with sufficient resources, do much to improve existing services throughout the country. The following paragraphs contain a short summary of the Committee's views regarding the most important issues involved in the improvement of library services.

A. Interlibrary Cooperation

The three largest libraries in this country, the Library of Congress, the Harvard University Library, and the New York Public Library, have long ago given up any hope of collecting all materials necessary for research. Considerable evidence exists that over the past 150 years, libraries in this country have been doubling in size every 16 years. This geometrical progression raises great problems requiring that attention be given to the various technical proposals which have been made for reducing the bulk of this material and for simplifying the problem of storage and cataloguing. Pending the widespread adoption of really revolutionary technical aids, it will be necessary to make comprehensive arrangements for interlibrary cooperation.

There are two important problems

here. One, that of securing in this country at least one copy of all needed items. Various estimates have suggested that existing library holdings represent from one-third to one-half of all the books published. In other words, nowhere in this country are there copies of many millions of books, pamphlets, magazines, etc. The second part of the problem is securing enough copies of various titles so that they are strategically available to students and scholars. For some books, perhaps, one copy would be enough, for others, however, there should be copies distributed according to some cooperative plan.

The participating institutions would then be free to reduce the rest of their collections to what may be called "working libraries." Adequate data are not yet at hand for defining the limits of such working libraries in the various scientific fields, and analyses of the sort recently undertaken by the Association of Research Libraries are urgently needed. Studies made by this group of the use of chemical periodicals suggest that a reasonable working library covering over 90 percent of the ordinary chemical laboratory's needs could be maintained by purchasing less than half of the available periodical literature.

Interlibrary cooperative plans could take the form of agreements among all libraries whereby each would attempt to be inclusive in limited fields. This would involve union catalogues on a regional and national basis and smoothly organized transportation arrangements. The Committee recognizes that proposals such as this have been frequently under study by librarians and that there are many difficulties to be surmounted. Federal aid for the library system of the coun-

try might well have as its central object the strengthening of the Library of Congress so that it could foster programs of cooperation. Both the Library of Congress and the Army Medical Library occupy leading positions in their fields. Yet these two Government institutions still have to look to private sources for much of their support, especially for projects involving experimentation with new methods. Two foundations alone have contributed over half a million dollars to the Library of Congress in the past few years.

Before leaving the subject of libraries it may be well to draw attention to an acute though temporary problem brought on by the war. Few, if any, European scientific publications for the last 5 years have been reaching this country in adequate volume. Many important periodicals published are not represented at all and others are available in only a few libraries and in broken sets. The end of the war will not bring about a solution to the problem since much of this material was published in reduced editions because of wartime restrictions on paper and printing. Furthermore, much existing material has been destroyed by enemy action. If American libraries are not to show serious gaps in their collections of important foreign periodical literature, it will be necessary to provide funds for reproducing much of this material. The funds necessary for such a project are entirely beyond private resources, and it is proposed, therefore, that the Government undertake to fill this need.

B. Abstracting and Translating Services

One of the most useful tools whereby the scientist is enabled to keep up with the flow of publication is the collection of abstracts published in several different fields. Their publication is extremely laborious and expensive, involving, as it does, the reading, summarizing, editing, and printing of all scientific articles published from month to month. At the present time much of this labor is donated by scientists who would rather sacrifice some of their research time than be deprived of this useful service. The existence of these publications is a precarious one and financial deficits are frequently encountered. Since the start of the war, many continental publications have not been available for abstracting and a large number of American and British papers have been withheld for security reasons.

If, as seems likely, it will not be possible to get this accumulated supply of papers abstracted by persons working without compensation, substantial costs may be expected.

With the advent of Russia as an important contributor to science, the problem of translating services becomes acute. Hitherto, most American scientists have been able to read the languages in which most scientific work was published. Unfortunately, there is little likelihood that many of the present generation will learn Russian in the near future. Translation and republication of important Russian works would, therefore, appear to be necessary and is likely to be very expensive. A study of the problem insofar as it concerns biology is now under way by the editors of *Biological Abstracts,* and within a period of several months it should be possible to give a reasonably accurate statement of the cost involved. Since such work would benefit not only science generally in the United States but would very likely promote the use of English in other countries, it seems proper to recommend that the United States Government consider methods by which the cost of such work could be met.

C. Bibliographic and Reference Services

The rapid accumulation of scientific publication continually increases the problem of keeping up with advances even within a single field. Every year earlier work is obscured by the mass of contemporary publication. Frequently discoveries have been published in the past which were neglected because their ultimate significance was not apparent in time. The task of bringing these past discoveries to bear on present problems is a difficult one. Searching the indices of a hundred different periodicals for the past several decades is an almost insuperable task. In some fields, notably medicine, consolidated and cumulative indices are available; in others the abstracting sources are an aid. For various reasons, however, these mechanisms are not wholly satisfactory. For example, the index catalogue of the Surgeon General's Library, which is the only publication attempting to gather together all medical publication in a single cumulative index, is up to date for only one letter of the alphabet in any 1 year. Few other fields, however, can boast of any cumulative catalogue of periodical articles.

Another attempt to meet the need is made by the reference services maintained as a part of library service. In theory, these organizations supply lists of publications bearing on particular subjects. In several fields an adequate job may be done

120

insofar as books or monographs are concerned, but rarely is sufficient attention given to periodical articles, which are far more important to the scientist. Furthermore, these services are frequently restricted in practice to older workers of distinguished reputation. The young man who wishes to be informed about past work is frequently compelled to divert much energy which could be better spent in his laboratory.

It seems probable that use of cataloguing and sorting devices now available in the form of business machines and the use of microfilm technique might go far to improve present methods of searching the literature and making bibliographies. Other technical advances may be expected further to simplify the problem. Adequate utilization of technical advance, however, would mean reclassifying all scientific literature for at least the past several decades. In the future this problem could be met by arranging for classification of every article prior to publication according to some prearranged system.

Again the Committee wishes to emphasize that it is not equipped to make specific recommendations in regard to technical library practice. It merely wishes to call attention to the existence of problems which, because of their magnitude and the large measure of centralization necessary for solution, appear to be proper subjects of Federal concern.

Analysis of University Research Expenditures

Research Expenditures in a Large Sample of American Colleges and Universities

To obtain factual information concerning research expenditures in 1939–40 and an estimate of postwar needs, questionnaires were sent to the 315 colleges and universities accredited by the Association of American Universities. Replies from 188 institutions were received, giving an over-all coverage of 60 percent. The coverage among the larger institutions was higher (over 80 percent) than that among the smaller schools. Of the 188 colleges reporting, 125 have organized research programs; the remaining 63 have not conducted research.

Table I summarizes the returns from the 125 institutions supporting research.

Table I

Expenditures and Needs of University Research

Natural Sciences and Engineering

	Operating expenditures [1]		Needs for post-war capital facilities [2]
	1939–40	Postwar projects	
10 largest institutions	$9,280,000	$16,342,000	$36,105,000
25 next in size	7,340,000	19,948,000	49,854,000
Totals, 125 research institutions	21,843,000	[3]47,716,000	108,290,000
Estimated totals, 150 research institutions [4]	26,213,000	57,260,000	129,949,000

[1] This category includes all items of research expense exclusive of buildings and items of major capital research equipment.

[2] These figures are estimates by the universities of their needs for capital facilities, including items of major capital research equipment and general laboratory facilities, without which the postwar research projects envisaged could not be carried on adequately.

[3] Of the $26,000,000 over and above prewar expenditures that would be required to finance these projects, the universities estimate that they will be able to raise only $7,000,000 through "normal" channels.

[4] It was estimated that, of the 315 colleges and universities accredited by the Association of American Universities, 150 have organized research programs in the natural sciences and engineering. Returns were received from 125 of these institutions. We have estimated the totals for 150 research universities by adding 20 percent to each category.

There are about 150 colleges and universities in the United States that have organized research programs in the natural sciences and engineering. We have estimated that research expenditures in these institutions amounted to $26,000,000 in 1939–40. Estimates of postwar projects call for annual expenditures of $57,000,000. Although this must be regarded as a very rough approximation, internal evidence from the returns, and amplifying statements accompanying many of them, suggest that if adequate funds and personnel were available the universities would be able to carry out projects of this magnitude. Many of the universities, for example, have prepared careful lists of important projects that their staffs wish to undertake after the war.

Postwar estimates for capital facilities were aggregated. They total $130,000,000, but most of them include facilities used jointly for teaching and research.

Some of the comments which accompanied replies to the questionnaires are given below:

The great discrepancy between the financial figures for 1939 and for our ideal after the war is easily explainable—and not on the basis that we are hoping to get some money from the Government. There is just one way that scientific research can prosper and grow apace here, and that is for promising young men to have the time to do it. Therefore, we are certain that if funds became available either from outside or some gift to the college about which we know nothing at the present time, we should like to use them for a considerable enlargement of our staffs, which would mean that the men promising in research might carry only a half-time teaching load. This we should hope in its totality would amount to the full time of two or three extra men in each department.

* * *

The estimates which we have made for the future represent somewhat large increases over our pre-war expenditures, but they are quite in line with the funds which we have been expending for research in the war years.

* * *

Our estimate of postwar needs for research are based upon definite information concerning the research which our departments in the sciences feel they are capable of undertaking and which they want very much to undertake as soon as funds are available.

* * *

Although the administration and faculty are giving greater recognition to the needs of such research, it is apparent that the state will hardly rise to the occasion with adequate appropriations. The university is, therefore, in need of considerable support for its research program in the postwar period.

Research in Small Colleges

Quite interesting reports were received from small colleges on their attitudes toward scientific research and their desire to encourage it. One small but well-endowed college in the East stated that its primary objective is a high quality of undergraduate teaching, and that it considers continued scholarly interest in research essential in order to obtain this objective. It has done so, in general, by four different means: (1) Grants-in-aid from a special fund. (2) Leaves of absence to permit faculty members to work at other institutions. (3) The maintenance of a well-equipped machine shop and carpenter shop with expert assistance for the sole use of the science departments. (4) A liberal purchase policy for instruments that can be used for both instruction and research. This resulted in a well-equipped electronics laboratory at the beginning of the war, which was put to valuable use at once when war broke out.

The type of policy outlined is much more liberal in its encouragement of research than was found in most of the smaller colleges. It does, however, represent the aims of a considerable number of them, and the adoption of similar policies by others certainly deserves encouragement.

Nonresearch Institutions

The returns from the 63 colleges that do not have organized research programs were mostly in the form of letters; few attempted to fill out the questionnaires. The replies are of some interest, however, and they suggest that the small liberal arts colleges fall into two definite groups. Some of these schools view themselves as purely teaching institutions and have no interest in developing research programs. Furthermore, a number of them are strongly opposed to Federal subsidy. The following comment is typical:

In general it is my opinion that the Federal Government should not undertake to establish any far-reaching program for the support of research in either public or private colleges or universities. I do not believe such relations can be established and permanently maintained without involving political control, which has proved so disastrous in Germany and other totalitarian states.

On the other hand, many of the smaller colleges feel that they could make important contributions to research, if funds were available to them for that purpose. Of the 63 nonresearch institutions reporting, 25 made statements, of which the following are typical: •

With adequate funds, we could reasonably expect a much greater devotion to research than obtains at present, for the members of our staff have both the interest and the training requisite for fruitful work.

* * *

Personally, I believe that if funds were available it would contribute toward the advancement of science to have one or two members of the staff of a liberal arts college engaged in minor projects of research, such as could be carried on satisfactorily with the equipment that such a college has.

In developing a program of postwar Federal aid to scientific research, attention should be given to the potentialities of these schools. To the extent that our sample is representative, at least 40 percent of the small liberal arts colleges in this country are desirous of conducting research, and are prevented from doing so by lack of funds.

Research Expenditures in a Small Sample of Leading Universities, Industrial Research Laboratories and Nonprofit Science Institutes

To obtain a detailed picture of research expenditures in natural science departments, a special investigation was made in 13 leading universities and institutes of technology throughout the country. In each university, the president was requested to appoint a consultant in the natural sciences to cooperate in securing the necessary information. Inquiries were made with respect to the departments of chemistry, physics and biology, and of electrical, mechanical and chemical engineering. In those universities which had medical schools, similar inquiries were made in the departments of anatomy, physiology, biochemistry, and bacteriology. The relevant data were then compared with research expenditures in 10 of our leading industrial laboratories and in 7 nonprofit science institutes.

It should be remembered, in assessing the results, that the data were prepared under pressure of time. The accounting systems in different institutions differ widely; in some instances, detailed figures could be obtained with relative ease, while in others it was necessary to make some rather crude guesses.

University Research Expenditures

Time and funds for research varied substantially between departments in the same university and between universities. Analysis of these variations suggests that much remains to be done if the majority of natural scientists with research interests are to be given the opportunities for research that are available in the most favored departments. Engineering departments, in general, appear to have very meager funds for research, although there are a few notable exceptions. At least two of the engineering schools under consideration have undertaken far-reaching expansions in research activity since the year 1940—expansions not due to the war, and, indeed, impeded by the war. If these are carried through according to present plans, the postwar research picture in these institutions will be very different from that for 1939–40.

Extraordinary variations are shown in the extent to which direct research expenditures are met from outside sources — especially grants from industry or the foundations. Many departments draw more than half of their support from these outside sources and some of them get all their funds in this way.

Comparisons were made of research expenditures in various university departments, industrial laboratories and science institutes. Expenses were divided into professional salaries and direct operating expenses of research.[1] This latter category included expenditures for equipment, apparatus, technical and research as-

[1] See tables II-VI inclusive, columns 4 and 6; and table VII, columns 2 and 3.

sistance, publishing costs associated with research, etc. A calculation was then made of the amount of these direct expenses in relation to professional salaries. The minimum figure in any of the industrial laboratories or science institutes studied was about forty cents per dollar of professional salary; typical figures are near one dollar, and in certain cases the figure was more than two, or even three, dollars. The highest figures for the university departments—with one or two striking exceptions, such as the chemistry department in institution number 8—are approximately the same as the lowest figures for the industrial laboratories and science institutes. Typical figures for university departments are about twenty cents per dollar of salary, and often are considerably lower.

Although it is very difficult to judge, there appeared to be a correlation between the research contributions of a university department and the amount of research assistance made available to its professional workers. In the institutions and departments less adequately provided with such support there are many men with research ability whose productivity could be significantly increased by the provision of more adequate research funds. Such funds might be used to diminish heavy teaching loads, which leave many men with little time for research, and to provide essential apparatus and technical assistance.

Time Devoted to Teaching and Research

The universities and engineering schools included in this survey rank among the leading institutions of the country. In all of them research is fostered and encouraged, and is considered an important factor in aca-

demic promotion. However, the various institutions differ considerably in the relative emphasis given to teaching and research. In a general way the 13 institutions may be said to fall into two groups:

In group A, comprising institutions 1, 5, 8, 10, and 13, the teaching load is relatively light, but varies considerably among individuals. For some members of the staff it is moderately high, while for others it is only 2 or 3 hours a week. These institutions are also likely to have some research professors who do no teaching at all. Most members of the staff are expected to devote more than half of their time to research. Of the five institutions that fall in group A, four are privately endowed. Nos. 1 and 5 are large universities in which a great deal of research is being actively carried on in all departments. No. 10 is a similar medium-size institution. No. 8 is an institution devoted primarily to the natural sciences and engineering. One very large State university (No. 13) also appears to belong in this category, at least, as regards some of its science departments.

Group B (Nos. 2, 4, 9, 11, and 12) is made up of important State universities. The teaching load is considerably heavier in this group, averaging around 12 class-room hours per week. Most members of the staff, however, are able to devote about one-quarter of their time to research, and sometimes more. Research professorships involving little or no teaching are extremely rare in these universities.

Three private institutions (Nos. 3, 6, and 7) appear to lie somewhere between groups A and B in regard to the relative allocation of time between teaching and research. No. 3

126

Table II

Analysis of Research in Selected University Departments (1939-40)

Departments of Physics

1 Institution number[a]	2 Number of professional personnel	3 Number of technicians, secretaries, etc.	4 Salaries of professional personnel (thousands)	5 Total department budget (thousands)	6 Direct operating expenses of research[b] (thousands)	7 Ratio column 6 to column 4	8 Percent of research funds from non-university sources	9 Number of graduate students[c]	10 Degrees awarded[c] M. A.	10 Degrees awarded[c] Ph. D.
1	35	16	$106	$169	$20	0.19	29	57.5	5.2	7.6
2	26	4	41	54	7.5	.18	93	—	—	—
3	51	9	90	115	18	.20	47	37	3	6
4[d]	52	10	92	—	34.5	.38	20	65	12	5
5	37	7	104	171	[e]39	.38	7	100	10	6
6	56	—	148	245	27	.18	33	55	3	10
7	18.5	—	23	30	4	.17	100	14	3	2
8	39.5	10	85	141	41	.48	95	53	8	8
9	31	7	37	62	9	.24	0	50	7	2
10	33	10	54	80	—	—	5	35	3	6
11	18	3	27	35	2	.08	100	7	2	2
12	47	13	79	123	30	.38	0	72	4	8

[a] As follows:
1. Large private university.
2. Large State university.
3. Large private university associated with large State agricultural school.
4. Large State university.
5. Large private university.
6. Large private engineering school.
7. Medium size liberal arts university.
8. Medium size private engineering school.
9. Large State university.
10. Medium size private university.
11. Large State university.
12. Medium size State university.
13. Large State university.

[b] Includes expenditures for equipment, apparatus, technical and research assistance, publishing costs associated with research, field trips, expeditions, etc.
[c] Average for the 3 years ending 1939–40.
[d] Includes astronomy and physiological optics.
[e] After deducting $10,000 spent on cyclotron.

Table III

Analysis of Research in Selected University Departments (1939-40)

Departments of Chemistry

Institution number[1]	Number of professional personnel	Number of technicians, secretaries, etc.	Salaries of professional personnel (thousands)	Total department budget (thousands)	Direct operating expenses of research[2] (thousands)	Ratio column 6 to column 4	Percent of research funds from non-university sources	Number of graduate students[3]	Degrees awarded[3]	
									M. A.	Ph. D.
1	52	12	$103	$161	$23	0.22	60	133	13	25
2 [4]	95	17	151	181	34	.23	85	108	12	3
3	63	7	86	114	13	.15	50	60	9	12
4	109	12	121	171	73	.60	38	161	21	18
5	82	20	194	236	43	.22	53	162	33	19
6	75		173	231	20	.12	10	80	5	18
7	32	8	58	88	13	.22	10	20	1	5
8	59.5	9	76	160	70	.92	83	34	3	5
9	55	2	69	101	11	.16	60	75	15	10
10	54	14	65	80	14	.22		61	2	13
11	21	18	126	163			7	124	18	18
12	39	4	55	77	20	.37	60	38	5	5
13	48	15	109	208	32	.29	0	45	3	12

[1] For explanation of institutions, see footnote a, table II.
[2] Includes expenditures for equipment, apparatus, technical and research assistance, publishing costs associated with research, field trips, expeditions, etc.
[3] Average for the 3 years ending 1939–40.
[4] Includes both chemistry and chemical engineering.

Table IV

Analysis of Research in Selected University Departments (1939-40)

Departments of Biology (Zoology and Botany)

Several of these departments include physiology, biochemistry, and biophysics, but no medical school departments are included

1	2	3	4	5	6	7	8	9	10	
Institution number[1]	Number of professional personnel	Number of technicians, secretaries, etc.	Salaries of professional personnel (thousands)	Total department budget (thousands)	Direct operating expenses of research[2] (thousands)	Ratio column 6 to column 4	Percent of research funds from non-university sources	Number of graduate students[3]	Degrees awarded[3]	
									M. A.	Ph. D.
1	52	11	$127	$169.5	$38.5	0.30	1	58	6	12
2	56	12	96	123	20.7	.22	65	49	9	7
3a[4]	34	10	78	100	10.5	.10				
3b[4]	417	113	1,033	2,752	336	.33				
3a+3b	451	123	1,111	2,852	346.5	.31	11	305	36	39
4	61	25	126	164	13	.11	60	127	27	11
5	34	8	106	129	45	.42	23	131	26	10
6	20	2	61	79	8	.13	50	25	1	1
7	26	1	50	75	2	.04		31	6	4
8	36	2	80	113	28	.35	70	16	1	4
10	14	3	31	52	11.4	.37	13	10	2	6
11a[5]	60	6	103	128			6	65	16	14
11b[6]	85	20	137	205			18	75	14	13
12	25	4	39	64	13	.33	100	23		2
13	53	21	86	172	[6]39	.45	70	45	8	7

[1] For explanation of institutions, see footnote a, table II.
[2] Includes expenditures for equipment, apparatus, technical and research assistance, publishing costs associated with research, field trips, expeditions, etc.
[3] Average for the 3 years ending 1939-40.
[4] [4]a denotes biology department under faculty of arts and sciences, [4]b denotes biological work in agricultural college of same institution.

[5] [5]a denotes zoology and botany under faculty of arts and sciences. [5]b includes total of 4 departments in college of agriculture (bacteriology, genetics, plant pathology, economic entomology).
[6] This includes a special item of $24,000 for a large botanical field trip to South America.

Table V
Analysis of Research in Selected University Departments (1939-40)
Departments of Anatomy, Physiology and Bacteriology

Anatomy

Institution number[1]	Number of professional personnel	Number of technicians, secretaries, etc.	Salaries of professional personnel (thousands)	Total department budget (thousands)	Direct operating expenses of research[2] (thousands)	Ratio column 6 to column 4	Percent of research funds from non-university sources	Number of graduate students[3]	Degrees awarded[3] M. A.	Degrees awarded[3] Ph. D.
1	17	5	$46	$61	$16.7	0.38	1	7	0	2
2	17	5	38	50	2.5	.07	10	12	3	3
4	11	3	28	33	3	.11	0	---	1	0
5	25	19	67	145	27	.46	80	7	---	2
11	13	3	25	35	---	---	0	12	1	1

Physiology

Institution number[1]	Number of professional personnel	Number of technicians, secretaries, etc.	Salaries of professional personnel (thousands)	Total department budget (thousands)	Direct operating expenses of research[2] (thousands)	Ratio column 6 to column 4	Percent of research funds from non-university sources	Number of graduate students[3]	Degrees awarded[3] M. A.	Degrees awarded[3] Ph. D.
1	20	7	$62	$117	$22	0.35	10	38	5	8
2	17	9	46	76	30	.65	55	8	2	1
4	11	9	35	74	4	.11	0	9	1	1
5	16	9	58	100	40	.69	50	7	0	2

Bacteriology

Institution number[1]	Number of professional personnel	Number of technicians, secretaries, etc.	Salaries of professional personnel (thousands)	Total department budget (thousands)	Direct operating expenses of research[2] (thousands)	Ratio column 6 to column 4	Percent of research funds from non-university sources	Number of graduate students[3]	Degrees awarded[3] M. A.	Degrees awarded[3] Ph. D.
1	22	17	$56	$83	$26	0.46	10	20	2	3
2	13	4	31	39	---	---	---	12	9	3
4	32	2	42	50	5	.12	85	57	7	2
10[4]	6	9	21	36	7.2	.34	25	10	2	1
11a	7	2	13	16	---	---	---	---	---	---

[1] For explanation of institutions, see footnote a, table II.
[2] Includes expenditures for equipment, apparatus, technical and research assistance, publishing costs associated with research, field trips, expeditions, etc.
[3] Average for the 3 years ending 1939-40.
[4] Department of Pharmacology which was regarded as comparable with respect to research expenditures.

Table VI

Analysis of Research in Selected University Departments (1939-40)

Departments of Chemical Engineering and Electrical Engineering

Chemical Engineering

1	2	3	4	5	6	7	8	9	10 Degrees awarded[3]	
Institution number[1]	Number of professional personnel	Number of technicians, secretaries, etc.	Salaries of professional personnel (thousands)	Total department budget (thousands)	Direct operating expenses of research[2] (thousands)	Ratio column 6 to column 4	Percent of research funds from non-university sources	Number of graduate students[3]	M. A.	Ph. D.
3.........	9	1	$17	$23	$3	0.18	-----	8	2	3
4.........	17	1	22	28	6	.28	-----	27	11	3
5.........	13	3	57	83	14	.24	75	75	24	6
6.........	52	-----	97	152	7.5	.08	-----	80	13	9
9.........	14	2	26	36	2	.08	0	30	7	4
10.........	8	2	16	20	3.7	.23	-----	6	0	1
11.........	10	2	27	49	-----	-----	3	12	4	1

The report for Chemical Engineering in Institution No. 2 is included in the report for the Chemistry Department in table III.

Table VI, continued, and footnotes, p. 132.

131

Table VI—Continued

Electrical Engineering

1	2	3	4	5	6	7	8	9	10	
Institution number [1]	Number of professional personnel	Number of technicians, secretaries, etc.	Salaries of professional personnel (thousands)	Total department budget (thousands)	Direct operating expenses of research [2] (thousands)	Ratio column 6 to column 4	Percent of research funds from non-university sources	Number of graduate students [3]	Degrees awarded [3]	
									M. A.	Ph. D.
3........	21	2	$51	$60	$1.2	0.02	------	6	4	1
4........	15	4	39	45	3.4	.09	------	21	6	1
5........	13	3	44	54	.4	------	0	25	10	1
6........	61	------	175	302	53	.30	------	68	38	3
8........	12	0	20	24	3.5	.18	40	20	12	3
9........	12	2	26	32	2.5	.10	25	5	4	0
10........	6	3	22	32	5.2	.24	80	16	1	2
11........	13	6	44	60	------	------	0	10	3	1

[1] For explanation of institutions, see footnote a, table II.
[2] Includes expenditures for equipment, apparatus, technical and research assistance, publishing costs associated with research, field trips, expeditions, etc.

[3] Average for the 3 years ending 1939–40.

Table VII

Analysis of Returns on Questionnaires Sent to Industrial Laboratories and Nonprofit Science Institutes

[1939 data]

Industrial Laboratories

1 Reporting institutions	2 Salaries of professional personnel (thousands)	3 Direct operating expenses of research [1] (thousands)	4 Number of technical research personnel per professional individual	5 Ratio column 3 to column 2
1 [2]	$253	$124	0.24	0.48
2	1,225	1,250	.40	1.02
3	1,025	3,709	1.04	3.62
4	393	154	.39	.39
5	610	285	.50	.47
6	91	49	.29	.54
7	380	286	1.89	.76
8	46	102	.70	2.23
9	100	69	2.00	.69
10	80	107	.13	1.34

Nonprofit Science Institutes

1 [3]	$51	$23	0.76	0.45
2	518	557	2.18	1.07
3	73	32	.53	.44
4	29	36	.18	1.24
5	26	55	1.50	2.12
6	98	80	.53	.82
7	347	483	.76	1.39

[1] Includes expenditures for equipment, apparatus, technical and research assistance, publishing costs associated with research, field trips, expeditions, etc.

[2] 1 and 7 are large electrical companies; 2 is a communications company; 3 and 4 are oil companies; 5 is a large and 9 a small chemical concern; 6 is a meat-packing company; 8 is a glass company; and 10 is a large pharmaceutical firm.

[3] 1 and 2 are institutions for medical research; 3, 4, and 5 for biological research; 6 and 7 for research in the physical sciences.

is a large privately endowed university, associated with a large State agricultural school. No. 6 is an important engineering school. No. 7 is a medium-size liberal arts university.

In engineering departments, the teaching schedule is generally considerably heavier than in physics, chemistry and biology; often it runs to 18 class-room hours per week. In some institutions, however, the teaching schedule for engineers is no heavier than in other departments of the university.

In the medical sciences, teaching (prewar) was frequently concentrated in one 4-month term, during which time the teaching load was fairly heavy; but the remaining 8 months commonly involved little or no teaching. Some institutions deviate from this pattern, but, on the whole, faculty members in the medical sciences tend to have a large fraction of their time available for research.

In several institutions the amount of teaching done by men in the lower academic ranks was considerably higher than for the full and associate professors. In other cases, the amount of teaching was practically identical, regardless of rank. Often there were marked differences between one department and another in the same university. For instance, the chemistry department in one institution reported the regular teaching load in hours per week as: full professors, 3; associate professors, 8; assistant professors, 10; instructors, 12. This is an unusual amount of variation with rank. The biology department in the same institution reported a uniform figure for all academic ranks. In this particular institution the chemistry department appears to have been much more extensively endowed than the other science departments.

Report of the Committee on Discovery and Development of Scientific Talent

Table of Contents

LETTER OF TRANSMITTAL

June 4, 1945.

Dr. Vannevar Bush,
Director, Office of Scientific Research and Development,
1530 P Street NW., Washington, D. C.

Dear Dr. Bush:

To assist you in making recommendations in response to President Roosevelt's letter of November 17, 1944, you assigned consideration of the fourth question in that letter to the following-named committee:

Dr. Henry A. Barton, director, American Institute of Physics.

Dr. C. Lalor Burdick, special assistant to the president, E. I. du Pont de Nemours & Co.

Dr. James B. Conant, president, Harvard University.

Dr. Watson Davis, director, Science Service.

Dr. Robert E. Doherty, president, Carnegie Institute of Technology.

Dr. Paul E. Elicker, executive secretary, National Association of Secondary School Principals.

Mr. Farnham P. Griffiths, lawyer, San Francisco.

Dr. W. S. Hunter, professor of psychology, Brown University.

Dr. T. R. McConnell, dean, College of Science, Literature, and the Arts, University of Minnesota.

Mr. Henry Allen Moe, secretary general, John Simon Guggenheim Memorial Foundation.

Mr. Walter S. Rogers, director, Institute of Current World Affairs.

Dr. Harlow Shapley, director, Harvard College Observatory.

Dr. Hugh S. Taylor, dean of the graduate school, Princeton University.

Dr. E. B. Wilson, professor of vital statistics, Harvard University School of Public Health.

Mr. Henry Chauncey and Mr. Lawrence K. Frank are the committee's secretaries.

The committee held only two meetings, but there has been constant interchange of materials by mail and we have conferred frequently among ourselves and with others. The report herewith presented is a joint effort and it is agreed to, both as to content and form, by the whole committee. This result has been arrived at, not by compromise, but by study of the relevant facts in the light of the committee members' varied experience, and by discussion and agreement upon the conclusions to be drawn from those facts in the light of our experience.

Respectfully submitted,

HENRY ALLEN MOE,
Chairman, Committee on Discovery and
Development of Scientific Talent.

SUMMARY

To the question asked of you by President Roosevelt, "Can an effective program be proposed for discovering and developing scientific talent in American youth so that the continuing future of scientific research in this country may be assured on a level comparable to what has been done during the war?", your committee reports affirmatively, stating their considered judgment that an effective program of support from the Federal Government to that end can be organized and, indeed, must be organized in order to assure the continuation of scientific and technological training and research on a scale adequate to the needs of the Nation, in peace or war. There is a long history of support granted by the Federal Government for training and research and it is our judgment that that type of support needs to be, and can be successfully, extended to provide for those highly talented youth with scientific interests and ability who must be assisted else they will not be able to obtain the scientific and engineering training which they merit and which the good of the Nation requires that they obtain.

Our proposals to these ends have two phases:

I. *Long-term plans*, aimed at ensuring through the long future an adequate supply of scientists and engineers by discovering and developing scientific talent in American youth; and

II. *Plans for the immediate future,* aimed at making up, in part, the deficits in the ranks of scientists and engineers resulting from the war and the Nation's Selective Service policy.

I. Long-Term Plans

The Evidence for Our Conclusions

The intelligence of the citizenry is a national resource which transcends in importance all other natural resources. To be effective, that intelligence must be trained. The evidence shows that many young citizens of high intelligence fail to get the training of which they are capable. The reasons for that failure are chiefly economic and geographical and can be remedied.

Existing provisions, by scholarships and fellowships, are inadequate to meet the needs of this group, nor will State, local, and private plans for such assistance, which are now under discussion, be adequate. Our plans, simply, are plans—as respects science and engineering—to train for the national welfare the highest ability of the youth of the Nation without regard to where it was born and raised and without regard to the size of the family income. Much of our evidence and many of our conclusions on that evidence are applicable to fields other than science and engineering; but our plans, naturally, do not go beyond our mandate to make effective plans for the discovery and development of scientific talent in American youth.

The Necessary and Desirable Extent of the Proposed Plans

Throughout our deliberations, we have had it in mind that, by scholarships and fellowships and monetary and other rewards in disproportionate amounts, too large a percentage of the Nation's high ability might be drawn into science with a result highly detrimental to the Nation and to science. Plans for the discovery and development of scientific talent must be related to the other needs of society for high ability. Since there never is enough ability at high levels to satisfy all the needs of our complex civilization for such ability, we would not seek to draw into science any more of it than science's proportionate share. In that spirit of reasonableness our plans are:

We recommend that each year 6,000 4-year scholarships be awarded to enable youth of scientific promise to work for bachelor's degrees in scientific and technological fields. We recommend further that 300 3-year fellowships be awarded each year to enable the recipients to obtain advanced training leading to doctoral degrees in science and technology. The maximum total of Scholars, if and when the plan is in full operation, would be 24,000 and the maximum total number of Fellows would be 900. Maximum annual costs if the plan is to be realized fully may reach, after the fourth year of operation, $29,000,000.

Outline of the Plan and of the Means for Achieving It

The Scholars should be chosen by State committees of selection and the Fellows by a national committee of selection. The Scholars shall be eligible for the fellowships but the fellowships shall also be open to other qualified students.

We recommend that, for the Scholars, the scale of support should be that provided by the GI Bill of Rights for veterans, namely up to $500 annually for tuition and other fees, plus $50 monthly for personal support if single, and $75 monthly if married. For the Fellows, there should be an allowance up to $500 for tuition and other fees and up to $100 monthly for personal support.

The Scholars and Fellows should be chosen solely on the basis of merit. without regard to sex, color, race, creed, or need.

All those who receive benefits under this plan, both Scholars and Fellows, should be enrolled in a National Science Reserve and be liable to call into the service of the Federal Government, in connection with scientific or technical work in time of war or other national emergency declared by Congress or proclaimed by the President. Thus, in addition to the general benefits to the Nation by reason of the addition to its trained ranks of such a corps of scientific workers, there would be a definite benefit to the Nation in having these scientific workers on call in national emergencies. Evidence presented to the committee shows that, if such a science reserve had been in existence

138

in 1940 and had included many of the best scientists, the mobilization of scientific men before Pearl Harbor would have been much more rapid and effective than it was possible to make it.

We believe that the obligation undertaken by the recipients of National Science Reserve scholarships and fellowships would constitute a real *quid pro quo* and that the Federal Government would be well advised to invest the money involved even if the benefits to the Nation were thought of solely —which they are not—in terms of national preparedness.

As merit should be the sole basis of selecting the Scholars and Fellows, likewise merit should be the sole basis of their continuing to hold their scholarships and fellowships—4 years for the Scholars and a maximum of 3 years for the Fellows. Unless the Scholars and Fellows maintain good behavior, good health, and scientific progress in the top quarter of their classes, the assistance they are receiving should be terminated.

The quotas of scholarships to be awarded by the State (and Territorial) committees of selection should be determined by the number of their secondary school graduates of the previous year as related to the national total of such graduates. That is, the national total of 6,000 scholarships would be prorated to the States in the same proportion as their high school graduates bear to the whole national total of such graduates.

We recommend that the establishment of the organization to operate the plan and its supervision be entrusted to the National Academy of Sciences— the top scientific organization of the country and the one which, through the years since its establishment in 1863, has shown itself to have the knowledge, integrity, ability to withstand pressures, and concern for the national welfare, which will be required.

II. Plans for the Immediate Future

Because Selective Service policies have not taken account of the Nation's vital needs for scientists and engineers, the training of men in the fields of science and technology during the war has almost completely stopped. Because of these stoppages, not until at least 6 years after the war will scientists trained for research emerge from the graduate schools in any significant quantities. Consequently, there is an accumulating deficit in the number of trained research scientists and that deficit will continue for a number of years.

The deficits of bachelor's degrees in science and technology are already probably about 150,000.

The deficits of scientific doctoral degrees—that is, of young scholars trained to the point where they are capable of original work—has been estimated, for the period 1941 projected to 1955, to be more than 16,000.

All patriotic citizens who are informed about these matters agree that, for military security, good public health, full employment, and a higher standard of living after the war, these deficits are very serious. Neither our allies, nor our enemies, permitted any such deficits to develop but on the contrary maintained or increased national programs for the training of scientists and engineers.

The feasible remedies in the situation, as we find it now, appear to us to be these:

1. We recommend that the Research Board for National Security and the Army and Navy find men who, before their induction and during their service, have shown promise of scientific ability and that they be ordered, by name, to duty in the United States as students for training in science and engineering of a grade and quality available to civilians in peacetime. This should be adopted as the considered policy of the armed services and no desire of a commanding officer to retain a potential scientist for his usefulness on the spot should be allowed to interfere with the operation of the policy.

Merit should be the sole basis for the selection of these students and merit alone should determine the number selected. We think that probably the total would be no more than 100,000 and that number, following VE-day, could not be militarily significant. For building up the Nation's scientific strength, however, that number would be very significant. If well selected on their merits as students of science, these men would constitute the premium crop of future scientists and we know that the future of our country in peace and war depends on that premium crop.

Under this proposed plan, be it noted, there would be no disruption of plans already made for the discharge of soldiers from the Army. While students, their discharges would occur in accordance with the already established rating scale. It would not do to propose that such a plan should be done on a volunteer basis—that is, that personnel of the Army and Navy should request orders to duty as students. It would not do because many of the best of them probably would not request such orders, from feelings that they would not wish to be put in the position of seeming to shirk their full patriotic duty.

2. The Army has made plans for setting up in foreign countries, when and where the military situation permits, courses of study for soldiers, including courses in science and technology. These plans are all to the good. The further important thing to ensure is that the courses shall be the best and most up-to-date that can be given. Unless it is to do a disservice to the soldiers taking its training, the Armed Forces Institute must be prepared with instruction that is wholly up to date in its higher levels; but the fact of the matter remains that since the Massachusetts Institute of Technology, the California Institute of Technology, the Ryerson Laboratory of the University of Chicago, and others, cannot be moved abroad, the plan for Army universities must be supplemented by what we have suggested in our first proposal above.

3. Public Law 346, Seventy-eighth Congress, commonly known as the GI Bill of Rights, provides for the education of veterans of this war under certain conditions, at the expense of the Federal Government. Among the returning soldiers and sailors will be many with marked scientific talent which should be developed, through further education, for the national good. However, the 1 year of education which the law provides for essentially all veterans clearly will not be enough to train a scientist nor in most instances to complete training begun prior to entry into the armed forces. The law makes the amount of education beyond 1 year at Government expense depend on length of service rather than on ability to profit from the education.

It appears to us that our mandate to set up an effective plan for discovering and developing scientific talent must take into account the scientific potentialities among the 10,000,000 young Americans now in the armed forces. To this end, it is recommended that the Veterans' Administration set up an adequate counseling service for those veterans of marked scientific talent and that a complete scientific education at Government expense be provided for a group of them selected on the basis of merit and irrespective of the length of their military service. Here, again, we believe it best to recommend that standards of scientific ability be the limiting factors rather than to suggest that definite numbers be selected for training. Amendment of the GI Bill of Rights, to make that law an instrument for the amelioration of the deficits of scientists resulting from the war and Selective Service policy, seems to us essential for the safety and continued prosperity of the Nation.

The adequate handling of the education of the scientific and technological talent now under arms will be a primary test of the effectiveness of the Government in meeting the whole problem to which we have been asked to direct our attention. The future scientific and technical leaders in the United States are now largely in military service. Unless exceptional steps are taken to recruit and train talent from the armed services at or before the close of the war, the future will find this country seriously handicapped for scientific and technological leadership. In peace or war, the handicap might prove fatal to our standards of living and to our way of life.

PREFACE

You asked us to advise you upon the fourth question of President Roosevelt's letter to you concerning future scientific development in the United States. The question is:

Can an effective program be proposed for discovering and developing scientific talent in American youth so that the continuing future of scientific research in this country may be assured on a level comparable to what has been done during the war?

In our judgment the answer to the question is in all respects in the affirmative. We conclude also that the program envisaged by the question is both necessary and desirable. The difficult questions are upon the necessary and desirable extent of such a program and upon the best means for its accomplishment.

Our report, accordingly, will be under three heads: the necessity, the extent and the means for making the envisaged program effective.

There are, however, some general considerations which we deem it well to place before you prior to proceeding to the body of this report.

President Roosevelt's letter to you looks toward a science that will be a decisive element in the national welfare in peace as it has been in war. He said, "New frontiers of the mind are before us, and if they are pioneered with the same vision, boldness, and drive with which we have waged this war we can create a fuller and more fruitful employment and a fuller and a more fruitful life." It is clear that the letter refers to science as the word is commonly understood, or, more technically described, to science now within the purview of the National Academy of Sciences, that is, to mathematics, the physical and biological sciences including psychology, geology, geography and anthropology and their engineering, industrial, agricultural and medical applications. To science in this sense, therefore, the recommendations in this report will be limited.

The statesmanship of science, however, requires that science be concerned with more than science. Science can only be an effective element in the national welfare as a member of a team, whether the condition be peace or war.

As citizens, as good citizens, we therefore think that we must have in mind while examining the question before us—the discovery and development of scientific talent—the needs of the whole national welfare. We could not suggest to you a program which would syphon into science and technology a disproportionately large share of the Nation's highest abilities, without doing harm to the Nation, nor, indeed, without crippling science. The very fruits of science become available only through enterprise, industry and wisdom on the part of others as well as scientists. Science cannot live by and unto itself alone.

This is not an idle fancy. Germany and Japan show us that it is not. They had fine science; but because they did not have governments "of the people, by the people and for the people" the world is now at war. This is not to say that science is responsible: it is to say, however, that, except as a member of a larger team, science is of limited value to the national welfare.

The uses to which high ability in youth can be put are various and, to a large extent, are determined by social pressures and rewards. When aided by selective devices for picking out scientifically talented youth, it is clear that large sums of money for scholarships and fellowships and monetary and other rewards in disproportionate amounts might draw into science too large a percentage of the Nation's high ability, with a result highly detrimental to the Nation and to science. Plans for the discovery and development of scientific talent must be related to the other needs of society for high ability: science, in the words of the man in the street, must not, and must not try to, hog it all. This is our deep conviction, and therefore the plans that we shall propose herein will endeavor to relate the need of the Nation for science to the needs of the Nation for high-grade trained minds in other fields. There is never enough ability at high levels to satisfy all the needs of the Nation; we would not seek to draw into science any more of it than science's proportionate share.

Through all ages of civilization far-seeing men and women and governments have been concerned with the necessity of providing for the leadership of the future, as one essential factor in the survival, or progress, of civilization. Provision for the leadership of the future is necessary because high ability, adventurous talent, is not born only into families that can pay for its development. It is a fact that a large proportion of the world's best brains and finest spirits have attained or accelerated their development through outside support, of the type that we should call scholarship or fellowship assistance. This is a profound social fact: a large part of the world's leaders in science and other fields of scholarship, in the creative arts, and even in public affairs, has required a financial leg up, while working toward leadership.

Upon any study of the history of the development of leadership we may be reasonably sure that a large part of the men and women who in future will lead us in all walks of life will need extraordinary boosts up the ladder at some stages of their careers—boosts provided by individuals, institutions, and governmental agencies, on the basis of a showing of very special ability —in the form of scholarships, fellowships, and grants-in-aid.

No nation has ever done as well as we have in equalizing educational opportunity, nor, probably, in giving the most adequate opportunity to the best; but it can easily be shown, and we shall show it, that we could do better. And we also shall show how we as a nation can do better.

Why we as a nation should be concerned to do better appears in the following statement by Dr. Robert Gordon Sproul, President of the University of California—a statement of such cogency and sound common sense that we are glad to adopt it as our own:

One of the major responsibilities of the university of the future, is to see that the money it spends * * * goes toward the education of the most worthy candidates in each generation. The intelligence of the citizenry of a nation is a natural resource which transcends in importance all other natural resources * * * One may condone the waste of many natural resources on the ground that science will some day discover a substitute that is just as good. But intelligence is quite unique, and though science search diligently it will never find a substitute for it, nor will the war lords.

Universities * * * are conservators of the above-average intelligence of the nation * * * Every conservation program must proceed along two lines: it must safeguard the known reserves of a given resource, and it must also, through exploration and every other means, make a determined effort to ascertain accurately the further supplies of that resource.

We do not know how much intelligence the citizenry of this Nation is capable of producing. We pay little attention to intelligence unless it forces itself to the surface and trickles into a college or university by force of gravity. If it happens to come to the surface in the backwoods area or a rural district, where the process of trickling down to college is made difficult by distance and by lack of funds, the chances are that the trickle will sink into the earth again, "unwept, unhonored, and unsung"—unless, of course, it happens to be one of the fastest running, highest-jumping, or quickest trickles on the track, court, or gridiron.

Across the continent from Dr. Sproul, Dr. James B. Conant, President of Harvard University and a member of this Committee, coming at the question from another direction, has made a statement to like effect which his colleagues of the committee would adopt as their own:

* * * in every section of the entire area where the word science may properly be applied, the limiting factor is a human one. We shall have rapid or slow advance in this direction or in that depending on the number of really first-class men who are engaged in the work in question. If I have learned anything from my experience in Washington as chairman of the National Defense Research Committee, it is that ten second-rate men are no substitute for one first-class man. It is no use pouring second-class men on a problem, even if you are under the greatest pressure for a solution; second-class men often do more harm than good. So in the last analysis, the future of science in this country will be determined by our basic educational policy.

And finally we would quote the Board of Regents of the State of New York who recently declared:

The need is imperative for enrolling the ablest young people of the State in institutions of higher education. This proposal is defensible not in terms of the desire of the colleges to obtain students. Fundamentally, the case rests on the need of any State to bring its best minds up to a high level of understanding and accomplishments.

This statement also we adopt as basic to our thinking.

The data which prove the truth of the quoted statements are well known and some parts of them are set forth in a Appendix A attached to this report. Here we simply give samples and it is to be noted that these samples apply not only to scientific ability in American youth but to ability generally:

An Indiana study published in 1922 showed:

If we compare the records made on our tests by the group of seniors representing the richest and the poorest homes, we find that there are proportionally more children possessing the highest grades of mental ability among the poorest class than among the wealthiest class, and more individuals with high average grades of intelligence among the wealthier than among the poorer group. The wealthiest group ranks high on central tendency. The poorest salaried group ranks low on central tendency and also

144

has a larger percentage of individuals possessing the lower grades of mental ability. But there are individuals in this class who obtain the highest intelligence rating made by high school seniors. * * *

It is still more significant that so many of this most superior group of high-school seniors will not attend college, while those with the most inferior grades of intelligence are planning to attend, in ever increasing numbers. Twenty-five percent of the brightest seniors found in the entire State said they were not planning to attend college at all, while 65 to 70 percent of the dullest seniors had definitely decided to go to college, most of them having already selected the college they expected to attend.

In Minnesota, it was shown that the probability of college attendance for a high school graduate of high college ability who is the son or daughter of a father employed in a professional or managerial group is several times greater than that for the son or daughter of a farmer or unskilled laborer. This study also showed that for every high school graduate who ranked in the upper 10 percent of his high school class and enrolled in college, another high school graduate who also ranked in the upper 10 percent did not enter college.

A Pennsylvania study showed that, in that State, dividing a sampling of the youth of approximately equal high ability into two socio-economic groups, 93 percent of children of the upper socio-economic group were graduated from high school and 57 percent attended college. But in the lower socio-economic group, only 72 percent of the children were graduated from high school and only 13 percent attended college.

As emphasized, this report is concerned with discovering and developing scientific talent, but in its proper setting and relationship to other needs for talent for the Nation's welfare. In the report we shall suggest, as befits our mandate, the appropriation of Federal funds to be applied only to the purpose of discovering and developing scientific talent; but, as we have pointed out, we recognize that there is need for the discovery and development of talent in all lines and we point out that most of the plans and procedures recommended herein for science are equally applicable to the discovery and development of talent in other fields.

What shall be done with Federal funds for the discovery and development of talent, scientific and other, in American youth is for the wisdom of Congress to determine. As taxpayers and as men concerned with the statesmanship of science, we have been deeply concerned with the question how plans for the use of Federal funds for scientific development may be set up so that Federal funds do not drive out of the picture funds from local governments, foundations, and private persons. We think that our proposals will minimize that effect, but, with proper candor, we do not think that our proposals will be completely effective to avoid what we do not want to happen. We think, however, that the Nation's need for more and better science is such that the risk must be accepted.

In this report, consonant with our mandate to make effective plans for the discovery and development of scientific talent in American youth, we recommend plans to assist able young men and women to carry their studies from the end of high school through the doctorate. Beyond that we do not go in our recommendations, not only because we do not think the word "youth" ought to be stretched to include men and women of post-doctoral

145

age, but also because your committees reporting upon other questions in President Roosevelt's letter are making recommendations for assistance to post-doctoral investigators. For our part, we are of opinion that the basic problem, at least for the next decade, will be to find more young talent and to give it a chance to develop into more first-rate investigators than we now have.

That is the problem at which this report aims. At present the opportunities for education beyond high school are accidental to too large an extent—determined by the accidents of geography and economic income. We seek, in this constitutional Republic, as respects science and engineering, to train for the national welfare the highest ability without regard to where it was born and reared and without regard to the size of the family income.

Part One

Long-term Plans

1. The Desirability of and Necessity for the Proposed Plans

We are convinced that there is no possibility that too much ability of the highest order can be discovered and developed: the needs of our complex social organization for brains and character at the highest level can never be surfeited. Moreover, it is appropriate to point out, when considering the need for scientific training, that the first-rate scientist and engineer cannot do his work effectively unless he has a few good ones in a secondary role at his disposal as assistants and sometimes a great many as hands and as instruments for the execution of his ideas.

We have only to look about us, from the point of view of citizens, to know that the current need for creative brains is not being met: there is too much wrong with the world and with our country to have doubt about that. As scholars and administrators of scholarly affairs we also know out of our own experiences that there is a deficiency in the supply of first-rate scientific workers. All of us know of problems in science whose solutions are urgently needed for individual and the collective welfare. The limiting factors, all along the line, are brains and character.

In Appendix A attached to this report, some startling figures are given as to the number of young people who drop out before completing high school. The country may be proud of the fact that 95 percent of boys and girls of fifth grade age are enrolled in our schools, but we cannot help being concerned with the fact that with each succeeding grade the percentage falls. For every 1,000 students in the fifth grade, 600 are lost to education before the end of high school has been reached, and all but 72 have ceased formal education before 4 years of college are completed. While this report is concerned primarily with methods of selecting and educating high school graduates at the college and higher levels, we cannot be unconcerned with the potential loss of talent which is inherent in the present situation in our primary and secondary schools.

The Nation's students may be diagrammed as a pyramid. At the base of the diagram are the students beginning the first grade. As we keep looking at this body of students, they drop out more and more rapidly and the sides of the diagram slope in sharply, making a pyramidal figure.

Students drop out of grade and high schools for a variety of reasons. The reasons which concern us in this report are only those which relate to the highly talented. Many of these individuals of great promise who are lost in the process are academic casualties undoubtedly to be charged against the inadequacy of the local secondary education available to them. Studies of the situation in different States show that the problem is by no means the same throughout the country. The figures that have been given above are the over-all figures for the Nation: in some States the loss is much less, in others much greater. Unless one were to believe — which we do not — that there is a corresponding difference in the distribution of native ability among the States, one cannot help reaching the conclusion that the differences reflect great variation in the quality of our secondary education.

It is not within our mandate to enter into the controversial subject of the way in which a larger amount of public funds should be expended on secondary education in those States where the amount now spent per pupil is very low, and it is surely no coincidence that it is in these very States that we find the losses, from the fifth grade on, to be the greatest. We would be remiss in our duty, however, if we did not point out that much remains to be done to make our educational system effective in developing the latent talent of the Nation by improving the quality of the secondary schools in many localities so that no boy or girl of talent and promise may be deprived of the proper high school education.

Among those who drop out before completing high school, both in the States which provide excellent public education and in those which are less advanced in this respect, there undoubtedly are some at least who have potentialities for becoming first-rate scientists and engineers. The early discovery of such individuals who have dropped out of the educational system obviously presents peculiar difficulties. The Committee suggests to employers, and to educational, scientific, and labor leaders that serious consideration be given to the problems involved in the discovery of such

individuals and in getting them back into educational institutions where their talents can be developed in spite of their lack of complete high school training.

Students drop out of high school, college, and graduate school, or do not get that far, for a variety of reasons. The reasons that concern us are only those which relate to the talented and they are (1) that they cannot afford to go, (2) that schools and colleges providing work of interest and up to the level of their abilities are not available locally, and (3) that business and industry recruit from among the ablest before they have finished the training of which they are capable.

These reasons apply generally, but they apply with particular force to science:

1. The educational road to becoming a high-grade scientist is long and expensive, and the families of many able students cannot afford to pay their way. It is of a length at least 6 years beyond high school and it is expensive because, as is evident, no large percentage of science students can get first-rate training in educational institutions while living at home.

2. Students of scientific capability are particularly vulnerable to bad or inadequate mathematical and scientific teaching in secondary school which fails to awaken their interest in science or to give them adequate instruction. Improvement in the teaching of science all along the line is imperative. To become a first-rate scientist it is necessary to get a good start early, and a good start early means good secondary school science teaching. No matter how gifted and capable a person may be, if he is not interested to finish secondary

school, or does not have the opportunity to complete secondary school, he cannot—as things are—go on to college and to graduate school.

3. Recruitment from among gifted students by business and industry likewise applies with particular force to science. A young man may well find the place in which eventually he will achieve high distinction in industry, following graduation from college, if his place ought to be, for example, in management or applied science. But if his place, considering his abilities, might be at the top in scientific research, he will be seriously handicapped if he stops his training without proceeding to the level represented by the doctorate. Industry and business cannot afford, as a long-term proposition, to recruit, prior to completion of training, those potential scientists who appear capable of contributing to fundamental advances or who should be teachers.

In the light of the studies made, having regard to the facts of the educational pyramid, it clearly is essential to provide for the early schooling of more able students in order that a large enough group will survive to become a larger quota of students of the highest ability at the apex of the pyramid. To increase this small group of exceptionally able men and women it is necessary to enlarge the number of students of high ability who go to college. This involves better high schools, provision for helping individual, talented students to finish high school (primarily, we conceive, responsibilities of every local community), and opportunities for more capable, promising high school students to go to college. Any other practice constitutes an indefensible and wasteful utilization of higher education and neglect of our human resources.

If we were all-knowing and all-wise we might, but we think probably not, write you a plan whereby there might be selected for training, which they otherwise would not get, those who, 20 years hence, would be scientific leaders and we might not bother about any lesser manifestations of scientific ability. But in the present state of knowledge a plan cannot be made which will select, and assist, only those young men and women who will give the top future leadership to science. To get top leadership there must be a relatively large base of high ability selected for development and then successive skimmings of the cream of ability at successive times and at higher levels. No one can select from the bottom those who will be the leaders at the top because unmeasured and unknown factors enter into scientific, or any, leadership. There are brains and character, strength and health, happiness and spiritual vitality, interest and motivation, and no one knows what else, that must needs enter into this supra-mathematical calculus.

We think we probably would not, even if we were all-wise and all-knowing, write you a plan whereby you would be assured of scientific leadership at one stroke. We think as we think because we are not interested in setting up an elect. We think it much the best plan, in this constitutional Republic, that opportunity be held out to all kinds and conditions of men whereby they can better themselves. This is the American way; this is the way the United States has become what it is. We think it very important that circumstances be such that there be no ceilings, other than ability itself, to intellectual ambition. We think it very important that every boy and girl shall know that, if he shows that he "has what it takes," the sky is the limit. Even if it be shown subsequently that he has not what it takes to go to the top, he will go further than he would otherwise go if there had been a ceiling beyond which he always knew he could not aspire.

By proceeding from point to point and taking stock on the way, by giving further opportunity to those who show themselves worthy of further opportunity, by giving the most opportunity to those who show themselves continually developing—this is the way we propose. This is the American way: a man works for what he gets.

2. The Desirable and Necessary Extent of the Proposed Long-Term Plans

As said in the general preamble to this report, we think that plans for the discovery and development of scientific talent should have a limit related to the needs of the Nation as a whole for trained talent in all activities that are necessary for the national welfare. We think, also as stated, that while we have no fears that too much top ability can be found and developed there is some danger that too many scientists of less than top ability may be trained, thereby debasing the currency of scientific training to the point where scientific careers may not look attractive either to the best or to the second best.

How to calculate the Nation's future needs for scientists, or to document fully a judgment upon the question, we confess we do not know. But we have some evidence to support what we, at any rate, regard as informed conclusions. This evidence is set forth in Appendix B attached hereto. In summary it shows the following facts germane to this report:

In the year 1941 there were conferred 53,534 undergraduate degrees in natural science and in technology.

In the last 6 years before the war, the average annual number of Ph.D. degrees conferred in natural science and technological fields was 1,649.

For some years to come, as pointed out elsewhere in this report, these numbers must be increased in an attempt to make up for the accumulated deficits in trained scientific and technological personnel caused by wartime interruptions to basic education and specialized training.

We have carefully studied data and indications concerning the Nation's future needs for scientists and technologists as a basis for determining the necessary and desirable extent of plans for discovering and developing scientific talent. We have concluded that the best that can be done is to make a practical, executive judgment after consideration of the material; and such a judgment leads us to propose that 6,000 science students annually be selected for assistance in obtaining the bachelor's degree. This number we judge to be not too large from any point of view or too small to be effective.

Similarly, making an executive judgment upon numbers of students proposed to be assisted annually to obtain doctoral degrees in science, we arrive at the figure 250, plus 50 for medical research doctorates unless your Committee upon the second question in President Roosevelt's letter makes a separate recommendation on fellowships in that field, which we understand is not their present intention. It is not intended that the

50 proposed medical-research pre-doctoral fellowships shall be administered nor allocated separately but simply that the recommended total number of predoctoral fellowships be increased to 300.

The number 250 is arrived at by considering, *inter alia*, that it would be 10 percent of the prewar average of science doctorates conferred, 165, plus a number endeavoring to make up some of the science doctoral deficit incurred during the war years when science students, practically, have been and are nonexistent. Our thinking concerning the added 50 medical research doctorates goes along the same lines.

These figures, we wish to emphasize, are not provable but equally we wish to emphasize they appear reasonable to us. It has been in our thinking throughout this report that we do not want to inflate or debase the currency of scientific training by artificially stimulating its issuance beyond the Nation's needs for such training.

Further, we desire to emphasize the point that, until we see the look of the postwar world, policies cannot be determined with finality. And, until policies can be determined, alternative plans, and sliding scales within those plans, are the only plans that make sense. We cannot, as we have said, guarantee that our figure of 6,000 assisted science students in each entering class and 250-300 assisted candidates for science doctorates a year are the correct figures for the needed result. We conclude, simply, that they are good figures with which to begin, always provided that they be not frozen and may be changed in the light of experience and as future demands for scientists and need for Federal assist-ance in training them may be shown.

Elsewhere in this report it is recommended that the administrative agency which may be charged with making our proposals operative be charged also with a continuing research function in which studies of opportunities for scientific and technological employment should have a major place.

When considering the question whether or not the group of undergraduates selected for training under the plan herein recommended be too large, it ought to be remembered that the majority will not go on to research careers but rather to various kinds of engineering practice, plant management activities and to many other kinds of practical work connected with industry and technological processes. For industries based on highly advanced scientific techniques which must be adapted constantly to new scientific discoveries, training in science is essential throughout the management, and while it cannot be said that a man, because he is a good scientist, is therefore a good manager for such a business, still without scientific training, he could hardly function at all. Moreover, for such a business a scientific training is, *qua* the business, probably as good a training as any other.

Furthermore, in reference to scientific training at the undergraduate level, we quote with approval a statement by a distinguished committee of English scholars, from social, humanistic and science fields, published by Nuffield College of the University of Oxford:

* * * We live in a world in which science lies at the very roots of community, and a mastery of scientific thinking grows more and more indispensable for the successful practice of the arts of life. The culture of the modern age, if it is to have

meaning, must be deeply imbued with scientific ways of thought. It must absorb science, without forsaking what is of value in the older ways or conduces to the understanding of those deeper problems which science by itself is impotent to answer. It is a question, not of substituting a scientific culture for that which has gone before, but of reaching a wider appreciation in which the sciences in their modern development fall into their due place * * *

3. The Recommended Long-Term Plan and Means for Achieving It

As stated in the preceding section, we propose that the number of undergraduate students of science and technology assisted under the plan shall be 6,000 annually and that the number of assisted doctoral students in the same fields shall be 250 or 300 annually. This would make the 4-year maximum total of undergraduates 24,000 and the 2- to 3-year maximum total of graduate students 900. Maximum annual costs, if the plan is to be realized fully, may reach, after the fourth year of operation, $29,000,000.

It is our idea that these highly selected students, if they proceed to doctorates, in many cases will be able to obtain that degree after 6 years of undergraduate and graduate work; but provision should be made for those who require 3, instead of 2, years of graduate work.

In this connection, we wish to emphasize the responsibility of educational institutions in this plan. Under the central purpose of the plan—to provide scientific training for students of superior ability and equal opportunity to all American youth to qualify in competition for such training— educational institutions will face the obligation of providing a training commensurate intellectually with the superior ability of this special group. The Committee believes that a pro-

gram which is appropriate for the rank and file of college students will not be appropriate for these, or other, highly selected individuals.

It appears to us that the scale of support for the undergraduate students selected under the proposed plan should be that provided by Congress under the so-called GI Bill of Rights, namely, tuition and other fees up to $500 annually and, for personal support, $50 a month during the months of each year when the Scholars actually are engaged in full-time study. Benefits under the plan should not be restricted to young and recent secondary school graduates but should be available also to those who, having worked in business and industry, desire to obtain scientific training at the college level. Such Scholars, if married, should receive, as also provided in the GI Bill of Rights, $75 monthly for support when engaged in full-time study.

Persons who receive benefits under the plan should be selected solely on the basis of merit, without regard to sex, race, color, or creed.

The question whether or not financial need should be considered as a factor in awarding benefits under the plan has been the subject of much study, consultation, and thoughtful consideration by the Committee. We conclude that need should not be a factor in the awards, for many reasons, among which are that, if need is to be considered, there would be required a means test of the parents, difficult if not impossible to administer with equity; those who receive benefits under the plan would be labeled as poor; and in cases where parents were not sympathetic to higher education their children might be cut off from the benefits of the plan.

Moreover, we consider that, apart from and in addition to the general benefits to the Nation flowing from the addition to its trained ranks of such a corps of scientific workers, there should be a definite and stated *quid pro quo* from the beneficiaries to the Nation. Hence, we propose that the beneficiaries under the plan should constitute a National Science Reserve, with definite and stated obligations to the Nation for scientific work similar to the obligations of members of the Army and Navy Reserves for service of the kind for which they have been prepared.

We suggest that recipients of undergraduate scholarships under the proposed plan be known as National Science Reserve Scholars and that recipients of predoctoral fellowships be called National Science Reserve Fellows.

The awards of Science Reserve Scholarships for college training for the bachelor's degree should be based upon tests of ability and aptitude to insure that the successful candidates will be oriented to scientific and technological pursuits. Moreover, acceptance of the Scholarships and Fellowships should be understood by the recipients as indicating intention to engage professionally in scientific and technological work but not as constituting an absolute obligation to do so.

We recommend that the recipient of a National Reserve Scholarship or Fellowship shall agree that, upon the completion of his undergraduate or graduate training, he shall be enrolled in the National Science Reserve and be liable to call into the service of the Federal Government, in connection with scientific or technical work in time of war or of a national emergency declared by Congress or proclaimed by the President

—the conditions of employment and the salary to be determined at that time by the President.

This call would be at the option of the Federal Government. It is contemplated that, in cases where men had not for years been engaged in scientific or technical activities, the Government probably would not exercise the right of call.

In addition to the binding obligation to serve the Government (if called) full-time in case of war or a national emergency, the members of the Reserve should pledge themselves to render assistance to the Government in time of peace, through service on advisory committees and on a consulting basis insofar as they are able to do so without gross interference with their professional work or the rendering of effective service to their employers.

We believe that the proposed National Science Reserve would be of real service to the Nation. Evidence presented to the Committee shows that, if such a science reserve had been in existence in 1940 and had included the best scientists, the mobilization of scientific and technical men to assist the Army and the Navy (directly and through OSRD), before Pearl Harbor, would have been more rapid and effective than it was possible to make it. We believe that the obligation undertaken by the recipients of National Science Reserve scholarships and fellowships would constitute a real *quid pro quo* and that the Federal Government would be well advised to invest the money involved even if the benefits to the Nation were thought of solely— which they are not—in terms of national preparedness.

The exact extent and duration of the obligation to serve, assumed by members of the National Science Reserve, of course, would be for the wisdom of Congress to determine in relation to the needs of the Nation and to the obligations of graduates of the Military and Naval Academies, of members of the Army and Naval Reserves and indeed of all citizens in time of war or other national emergency.

It is agreed by the Committee in respect to the administration of the National Science Reserve Scholarships that while the plan must be national in character, the principle of local administration must be recognized. The American scene which looks rather uniform from any one place has infinite variety and intense individuality at close range. This must be recognized.

Our plan for the selection of National Science Reserve Scholars is set forth tentatively in Appendix C attached hereto. In brief it is this:

The 6,000 proposed scholarships should be assigned to the States [1] on the basis of the number of their secondary school graduates of the previous year as related to the national total of such graduates.

On the basis of the 1939–40 figures, State quotas of scholars would be as stated in the table on the following page.

It is recommended that, for the National Science Reserve Scholars, the administrative organization, the bases of selection and the procedures be as follows in brief:

Proposals for a "National Scientific

Research Foundation" are under discussion by your Committees and among the proposed powers of such a foundation is power to contract with other agencies for the performance of functions within the scope of the foundation. It would be our recommendation that the foundation, or any similar organization which may be established, should make arrangements for choosing Scholars and Fellows under the proposed plan through the National Academy of Sciences, if that organization be willing to accept the responsibility. The operation of the plan, we recommend, should be entrusted to the Academy's operating agency, the National Research Council. More precise details of the National Academy's participation and the means by which it is suggested that operations be carried on are stated in Appendix C.

To ensure the fairest, most effective and most up-to-date methods of selection, advisory bodies expert in such matters must be set up. No existing national science organization has shown itself to be as well-equipped for such advisory functions —working both with nonmembers of the academy and with members— as the National Academy has shown itself to be through the years. A central administrative staff, chosen for ability and integrity, also must be set up.

As outlined in Appendix C, committees of selection would be set up in each State. These committees, it is suggested, should consist of five members, *to wit*: three scientists, one of whom should serve as chairman; one member of a college or university faculty, trained and experienced in the field of selection and guidance; and one representative of secondary

---

[1] It is intended that the proposed scholarships shall be available also to secondary school graduates in the Territories and Insular Possessions but we have not statistics relating to them comparable to those for the States and for the District of Columbia given on this page. Allocation of scholarships to the Territories and Insular Possessions, of course, would decrease the State quotas.

State	Secondary school graduates [1]	State quotas: scholarships
Alabama	16,222	80
Arizona	3,498	17
Arkansas	12,226	60
California	72,301	356
Colorado	11,900	59
Connecticut	17,614	87
Delaware	2,353	12
District of Columbia	5,278	26
Florida	12,666	62
Georgia	18,302	90
Idaho	6,815	34
Illinois	75,508	372
Indiana	37,470	184
Iowa	30,671	151
Kansas	23,326	115
Kentucky	17,675	87
Louisiana	17,405	86
Maine	8,485	42
Maryland	13,016	64
Massachusetts	46,830	231
Michigan	44,522	219
Minnesota	30,337	149
Mississippi	13,979	69
Missouri	33,343	164
Montana	6,617	33
Nebraska	17,970	88
Nevada	1,005	5
New Hampshire	4,670	23
New Jersey	39,973	197
New Mexico	3,745	18
New York	117,901	580
North Carolina	30,372	150
North Dakota	7,182	35
Ohio	73,616	362
Oklahoma	23,467	116
Oregon	13,002	64
Pennsylvania	99,351	489
Rhode Island	5,978	29
South Carolina	12,687	62
South Dakota	8,059	40
Tennessee	17,857	88
Texas	56,348	277
Utah	8,212	40
Vermont	3,130	15
Virginia	20,263	100
Washington	21,170	104
West Virginia	17,571	87
Wisconsin	33,464	165
Wyoming	3,213	16
Totals	1,218,545	5,999

[1] Public high school graduates 1939-40 plus 1/6 of the private and parochial secondary school enrollment. Statistics of State School Systems, 1939-40, 1941-42, Biennial Survey of Education, U. S. Office of Education.

education within the State, usually a secondary school principal or a high school supervisor in the State department of education. At least one of the scientists, it is suggested, should be from agriculture or from industry within the State.

The administrative staff of the national over-all organization, in cooperation with the advisory bodies, would prepare tests in accordance with the best thought upon such matters. These tests would be given to all applicants throughout the country and the test reports would be sent to the national organization for evaluation. The national staff also would collect other relevant data and judgments concerning each applicant and upon the basis of the tests and other material would certify to the State committees a number of qualified candidates equal to twice the State's quota—it being provided, however, that no applicants shall be certified who do not attain a certain minimum national standard. The dossiers of these candidates would be sent to the State committees and those committees would have the responsibility of making the final selections of the Scholars up to the number of the State's quota.

The machinery and procedures for administering these scholarships are outlined only in general terms at this time. It is clear that valid methods for selecting students of high ability are available in the experience of persons and organizations which have been working on this problem for many years. Doubtless better methods will be available in future and the methods adopted for the National Science Reserve Scholarships should be the best available at the time they are being used.

Concerning machinery for administration of the National Science Reserve Fellowships, we can be brief:

They should be administered nationally as the National Research Council Fellowships are administered. Whether or not a "National Scientific Research Foundation" or similar body be established, we judge that the National Research Council of the National Academy of Sciences would be the best agency to administer the proposed National Science Reserve Fellowships; for the National Research Council has shown that it has the know-how and integrity to administer well a Fellowship program on a national basis. It should be noted that we propose that the fellowships (as distinct from the scholarships) should be awarded on a national, not a State, basis.

For the National Scientific Reserve Fellowships as for the Scholarships, the sole basis of selection should be merit, without regard to sex, race, color, creed, or need.

It is recommended that fellowship (as distinct from scholarship) stipends should be fixed by the awarding agency on a scale up to $100 monthly, plus payments for tuition and other fees up to a maximum of $500 annually.

Throughout the whole plan, for both scholarships and fellowships, there should run an insistence upon high-grade work by the holder, otherwise the fellowship or scholarship shall be terminated by the awarding agency.

Scholarships shall be tenable for 4 academic years or the equivalent. Fellowships shall be tenable for the duration of graduate studies leading to the doctoral degree, up to a maximum of 3 academic years or the equivalent. Both shall be held upon the following conditions:

(a) Continuance of good health.

(b) Continuance of good behavior.

(c) Scientific progress at the level of the best 25 percent of former students in the scientific departments primarily concerned.

If a Scholar or Fellow drops out

for failure or other reasons, his scholarship or fellowship should lapse. Alternates should not be appointed.

The scholarships and fellowships should be valid for any college or university, within the territory of the United States, of the holder's choice, subject to the advice and consent of the awarding agency concerning relevant facilities for scientific work. With the consent of the awarding agency, a Scholar or Fellow may change the location of his work to another college, university or technical school which is judged to be better suited to his scientific development.

National Science Reserve Scholars should be eligible to appointment as National Science Reserve Fellows, but appointments to the fellowships should not be restricted to the National Science Reserve Scholars. The fellowships should be open to competition from all pre-doctoral science students.

It is recommended that the award of the scholarships and of the fellowships be commenced simultaneously, or approximately so, to the full annual number in each category.

The Committee recommends that the National Agency in charge of the scholarships and fellowships should carry on continuing research into methods of selection and continuing study of unfolding opportunities (and the reverse) for employment in science and technology.

Among effective means for the discovery and development of scientific talent in American youth are means for developing public interest in science. It will not be sufficient, if science is to remain healthy in root and branch, merely to develop a large number of scientists and to provide them with the financial support necessary for their investigations. There is also the necessity of creating a better understanding of the role and place of science in our national life, so that public approval and support for the future development of science will be forthcoming.

Part Two

Plans for the Near Future

The preceding sections of this report propose plans for the discovery and development of scientific talent in American youth as a long-term proposition. There is, however, an immediate and pressing problem which is a result of the war.

1. Deficits of Scientific and Technological Personnel Resulting from War and Selective Service Policies

The training of men in the fields of science and technology during the war has almost completely stopped. With the exception of some 2,400 men on the reserved list who have been taken from their studies for civilian war research, all physically fit graduate students have been taken into the armed forces. College students majoring in the sciences have also been taken into the armed forces. Those ready for college training in

the sciences have not been permitted to enter. Because of these curtailments, it will require at least 6 years after the war ends before scientists trained for research will emerge from the graduate schools in any significant quantity. Consequently there is an accumulating deficit in the number of trained research scientists. That deficit will continue for a number of years.

The deficits in science and technology students who, but for the war, would have been granted bachelor's degrees in these fields are probably already about 150,000.

The deficits, in science and technology, of doctoral degrees—that is, of young scholars trained to the point where they are capable of carrying on original work—have been calculated by the American Institute of Physics, as follows:

	Deficit accumulated 1941 through 1944	Estimated deficit 1945	Total 1941 through 1945	Probable deficits 1946 through 1955	Total probable deficit due to war 1941 through 1955
CHEMISTRY	240	550	790	4,460	5,250
ENGINEERING	148	82	230	730	960
GEOLOGY	63	50	113	317	430
MATHEMATICS	161	100	261	939	1,200
PHYSICS	251	160	411	1,589	2,000
PSYCHOLOGY	96	84	180	550	730
BIOLOGICAL SCIENCES	665	725	1,390	4,910	6,300
Totals	1,624	1,751	3,375	13,495	16,870

All patriotic citizens, who are well-informed on these matters, agree that, for military security, good public health, full employment and a higher standard of living after the war, these deficits are very serious.

In a recent radio address Dr. Arthur H. Compton, Professor of Physics in the University of Chicago and Nobel Prize winner, said:

It takes at least 6 years for a capable 18-year-old to train himself for effective scientific research. Even if we should start now to resume such training, it will thus be at least 6 years before a normal supply of young professionals will again be available to our laboratories. Can we afford to wait any longer?

Admiral J. A. Furer, Coordinator of Research and Development, United States Navy Department, has said:

I want to mention the great personal interest that the Secretary of War, Mr. Stimson, and the Secretary of the Navy, Mr. Forrestal, are taking in postwar military research. There is a growing belief that important as it may be to maintain after the war ground forces, air forces, and sea forces of a size commensurate with our national responsibilities, it may be even more important to keep the weapons and the material in general which we supply to these forces in step with the advances of science. Stocking our arsenals with the weapons of this war is no guarantee that we can win the next war with them. In fact, that may be the quickest way of losing the next war. It would be wiser to maintain arsenals of only modest size whether we are speaking of ships or guns or aircraft and to use the money saved thereby to continually replace the old things with the new creations of the research laboratory and of American inventive genius. Our industry should be kept alert to begin quickly the production of the vast quantities of materials needed when war threatens; and this readiness should concern itself especially with the new things. We hope for your aid in supporting this position among those who are engaged in research.

Dr. Charles L. Parsons, Secretary of the American Chemical Society, wrote President Roosevelt:

American technology has given birth to the greatest power of all time. Today, we are drying up prosperity at its source. Public opinion of the future will view with amazement the waste of scientists in World War II * * * Our children and our grandchildren will not forgive the loss of an entire generation of scientists.

Dr. Charles Allen Thomas, director of the Monsanto Chemical Company's research laboratories, declared:

Scientific suicide faces America unless immediate and adequate steps are taken to train replacements for technical men going into the armed services.

Statements of this type are fairly representative of the thinking of informed men in the armed services and in civilian life.

The situation, in brief, is that since the passage of the Selective Service Act in the autumn of 1940, there have been practically no students over 18, outside of students of medicine and engineering in Army and Navy programs, and a few 4-F's, who have followed an integrated scientific program in the United States. Neither our allies nor, so far as is known, our enemies have permitted any such condition to develop; but on the contrary have maintained or increased national programs for the training of scientists and technologists. It takes at least 6 years for a capable 18-year-old person to train himself for effective scientific research. Having regard to this long period of training and on the basis of prewar figures showing both the number of students of physical science in graduate schools and of doctoral degrees then conferred, the accumulating deficit of scientists has been calculated, with the results already presented. That these deficits

159

are a serious matter for the welfare of the Nation be the condition peace or war, is agreed. What are the feasible remedies?

Proposals to change the policy of draft boards to the end that students of science and technology shall not be drafted are too late. The damage has been done: these students already are in the Army and Navy, cut off from integrated scientific and technological training. Proposals for their early discharge from the Army and Navy are not feasible. The Army has made its plans for the discharge of personnel as soon as feasible in accordance with a rating scale conceded to be fair and reasonable from the standpoint of the individual GI—however it may disregard the risk to the Nation's scientific strength. Plans for the discovery and development of scientific talent in American youth who are in the Army and Navy must, to be practicable and reasonable, take account of the existing situation and of plans for demobilization already adopted.

Our proposals, in the situation as we find it, are these:

2. Plans for Integrated Scientific Training for Soldiers and Sailors

There should be prepared now lists of promising students of science and technology—students who before and after their entry into the armed services have shown high ability in these fields. Arrangements should be made now with the Army and the Navy whereby, now that it is militarily feasible, these talented students should be ordered to duty in the United States for fully independent, integrated scientific study of a grade available to civilians in peace times. This should be adopted as the considered policy of the armed services and no desire of a commanding officer to retain a potential scientist for his usefulness on the spot should be allowed to interfere with the operation of the policy.

It is recommended that this plan be carried out, not in terms of a stated number of young scientists, but rather that, now, centers of science and technology in the United States should be combed for information concerning those students who, prior to the war, had given evidence of high talent for science and technology; and that, as soon as militarily possible, these students by name, should be ordered to duty as students. Probably no more than 100,000 of the 10,000,000 men in the Army and Navy would be involved and now, following VE-day, that number could not be militarily significant. Likewise, we recommend that the armed services comb their records for men who, during the war, have given evidence of high talent for science and technology, and that they also be included in this plan.

It is recommended that the plan be not restricted to students at any particular level of studies, but rather that science students who have shown their abilities at all levels of studies, from college freshman to postdoctoral students, be included. It is also specifically recommended that former teachers of science in the armed forces be included in this plan.

The machinery for the discovery of the students under this plan, we venture to suggest, could best be set up within the Research Board for National Security.

Under this proposed plan, be it noted, there would be no disruption of plans already made for the discharge of soldiers from the Army;

while students, their discharges would occur in accordance with the already established rating scale. It would not do to propose that such a plan should be done on a volunteer basis—that is, that personnel of the Army and Navy should request orders to duty as students. It would not do because many of the best of them probably would elect to remain in the armed services, inspired by feelings that they would not wish to be put in the position of seeming to shirk their full patriotic duty.

Our recommendation is emphasized in the cases of men whose scientific training was well started before their induction, the more so the further that training had advanced. It is important to remember that the induction of many students in the critical science and technological fields was delayed and that under actual demobilizing plans they will consequently be among the last to return to civil life. A way must be found to insure the quick resumption of their training, composing, as they do, the recognized "premium crop" of science and technology.

The future of our country in peace and war depends on that premium crop.

3. The Importance of Quality of Instruction in "Army Universities" Abroad

The Army has made plans for setting up in foreign countries, when and where the military situation permits, courses of study for soldiers, including courses in science and technology. These plans are all to the good. The further important thing to ensure is that the courses shall be the best and most up-to-date that can be given, and shall include adequate laboratory work. You stated the issue in your letter of November 19, 1944, to General Frederick H. Osborn:

There have been in this country, by reason of war research, advances which will gradually permeate our entire industrial, scientific, and technical structure. Are the metallurgists now in the Army to return and find that they have studied alloys that are out-of-date? Are mechanical engineers to find that advanced thinking on gas turbines has outpaced those who have been at the front, and the new knowledge has not been extended to them? Are the large number of medical men in the field to have no direct contact until they return with those who have made more advance in medical research in the last few years than usually occurs in a decade?

It must be ensured that these questions can be answered in the negative. The Armed Forces Institute must be prepared with instruction that is wholly up-to-date in its higher levels; but the fact of the matter remains that since the Massachusetts Institute of Technology, the California Institute of Technology, the Ryerson Laboratory of the University of Chicago, and others, cannot be moved abroad, the plan for Army universities must be supplemented by what we have suggested in our first proposal above.

The Committee emphasizes that for men of scientific promise and ability there is special need that the Armed Forces Institute have its instruction modern, up-to-date, and of the best effectiveness. It is clear that there is a vast opportunity in this program for strengthening the technical work of the country by integrating the training given to soldiers possessing technical proficiencies with problems of modern industry and technology, especially for men who do not plan to go on to advanced scientific training. So far as possible, the universities and technical schools of the country doubtless would stand ready to cooperate with the Armed Forces Institute along these lines, if requested by the Army, by sending overseas instructors in technical and scientific subjects—instructors, who, fresh from war research, would be up-to-date. Technical branches of business and industry might well do the same.

4. The Place of the GI Bill of Rights in Ameliorating Scientific and Technological Deficits

Public Law 346, Seventy-eighth Congress, commonly known as the GI Bill of Rights, provides for the education of veterans of this war under certain conditions, at the expense of the Federal Government. Among the returning soldiers and sailors will be many with marked scientific talent which should be developed through further education, for the national good. However, the 1 year of education which the law provides for essentially all veterans clearly will not be enough to train a scientist nor in most instances to complete training begun prior to entry into the armed forces. The law makes the amount of education beyond 1 year at Government expense depend on length of service rather than on ability to profit from the education. It would seem to us that our mandate to set up an effective plan for discovering and developing scientific talent must take into account the scientific potentialities among the 10,000,000 young Americans now in the armed forces. Accordingly, it is recommended that:

(a) a special advisory committee of scientists be appointed to assist the administrators of the law to discover and direct the counseling of those veterans who have marked scientific talent;

(b) an adequate advising and counseling service be established in each State or region; and

(c) a complete scientific education at Government expense be provided for a group selected on the basis of the educational record of the first year (assured to all veterans) and such other tests as may be necessary—the length of this education to be determined, on the merits of each case, by the special advisory committee.

Under the suggested plans interested veterans while studying science for the first year, during which as veterans they are entitled to support from the Federal Government, would submit their records and take certain tests. Outstanding men and women of scientific talent would be selected—and we recommend a selective process as rigorous as that provided under our main plan—and be provided with funds at the rate prevailing in the GI Bill of Rights for completion of college courses in science, and also for graduate training to those possessing very high abilities.

We are informed that to some extent the proposals herein outlined could be put into effect under the present law by administrative action, and we venture to recommend such action to the extent allowable. We believe, however, that it would be advisable, in addition, to have new legislation authorizing the administrators of the law to select, as an estimate, possibly 5,000 veterans of each age group (i. e., those born in a given calendar year) for scientific education at the expense of the Federal Government (at the rates specified in present laws) irrespective of the length of their military service and up to a total of 6 years. Here, we think it sounder to recommend that standards of scientific and technological ability be the limiting factors, rather than to recommend that definite numbers of veterans be selected for training. In dealing with the veterans, for whom we think the best possible training should be offered, the only sound way for the administrators of the law to proceed is qualitatively, on the basis of assisting those who can maintain the highest standards, rather than on the basis of any quantitative estimates or fixed quotas.

It is not necessary to stress further that the proper handling of the reservoir of scientific talent now in the armed forces is of the first importance from the point of view of continuity in future supply of scientists. Not all of the scientific talent in the age groups here considered (those born in the years 1921 to 1928, roughly) is to be found in the armed forces, because some of the trained scientists among them have been kept at civilian tasks of utmost urgency for the war effort. However, such assignment to civilian status through Selective Service mechanism has been far from effective in the past year or two, and for those born later than 1924 (now 21 years of age or less), practically no exemptions from military service (except by reason of physical disability) have been allowed. Each year that the 18-year olds are called up for service in the armed forces a large portion of the potential scientific talent of that age group is cut off from adequate training. Among these younger men are those who will be the most promising candidates for further scientific education when demobilized; yet, because, under the provisions of the present law, the length of education depends on length of service, it will

be those young men who can have the least Government assistance. Amendment of the law to rectify this situation, at least insofar as future scientists are concerned, seems to us essential for the safety and continued prosperity of the Nation.

The relation of the proposed extensions of the provisions of the GI Bill of Rights to the long-term plan envisaged earlier in this report for the National Science Reserve is obvious. Those educated in science under the veterans' law for a period prolonged beyond the period to which as veterans they would be entitled should likewise be members of the National Science Reserve. The relation of the proposed extensions to our proposals for ordering members of the armed forces to duty as students likewise is obvious. That group would remain in the armed services only as long as, under actual plans for demobilization, they are required to remain. Thereafter, they would take up the educational benefits to which they will be entitled under the GI Bill of Rights, and under, we trust, our proposed extensions of benefits to the specially talented among them.

5. Duties of Schools, Colleges, Universities and Technical Schools to Returning Veterans

However, this is a problem not only for the Federal Government to solve, but also is one requiring that the States and the colleges, universities, and technical schools take leadership. We say emphatically to the colleges, and universities and technical schools that it is up to them to be extremely flexible and broadminded in handling the returning veteran. Unless they are willing and able to devise ways and means of developing in science those able veterans who do not meet the usual formal requirements, they will lose some of the best talent in the country. In particular, they must devise means of building on the basis of the very partial but highly detailed technical training that many of these men have received in service. Some of this large group of men, perhaps millions, who have learned about machines and electrical equipment can be further developed, for the well-being of the country, through special institutions or vocational schools. Moreover, from this group can be culled first-rate scientific talent, provided that the universities and technical schools do what they ought to do. The rigidity of academic institutions must not be permitted to drive away from training talented veterans.

We recognize a dilemma here: the scientific professions, including medicine and indeed all the learned professions as well, nowadays require, because of the complexities and vast extent of modern knowledge, both breadth and intensity in preparation. On the other hand, the generation with which we are concerned has already lost up to 5 years of educational time, and if the most ambitious among them are not to be repelled, ways must be found to shorten the period required for them to complete their formal education. It is a condition, not a theory, that confronts us and our judgment is that the Nation will lose much if our educational institutions do not recognize that many veterans will feel the need for making up lost time, and help them make it up. Otherwise, we are sure, a significant quantity of them will be lost to higher education.

Further, there is the problem of veterans needing to complete their secondary school training. Many of them, interested in completing their interrupted high school programs, will be deterred from doing so because, by reason of their greater age and maturity, they will be reluctant to go back to regular school classes with adolescents, to submit to the usual high school routines and requirements, and otherwise to live and associate with such youngsters. This situation must be met. A similar problem confronts many youths employed in war industries.

Provision for these "over-age" high school students is very necessary in postwar educational programs, especially for those who are not primarily concerned with vocational training, which apparently will be amply provided under present and proposed programs. Special provisions, such as those stated by the Regents of the State of New York, must be put into effect to make it attractive for able and promising youth to complete high school and thereby become eligible for college under one or more of the scholarship plans that will be available for talented high school graduates. Otherwise they will be lost to science and to higher education, generally.

In considering plans and programs for discovering and developing scientific talent in American youth, the needs of these particular groups must not be overlooked since they will include some of the potential leaders of the future, especially among the veterans who will have had war experience that has helped them to mature and develop. They must not be penalized for their priceless advantage, not now recognized in our regular educational arrangements.

The "Regents' Plan for Postwar Education in the State of New York" makes the following statement on this subject:

The men and women demobilized from the armed forces, together with workers of like age released from war industries, will include many thousands of persons whose educational career was interrupted below high school graduation. The military personnel will probably be granted scholarships large enough to take care of personal expenses. If offered a flexible program at the secondary level, with appropriate allowances for military experience and for work in the Armed Forces Institute, many of these veterans would fit into classroom, laboratory, and shop instruction. Others will be older and perhaps averse to receiving regular instruction in company with young pupils.

School authorities should make an inventory of all building, staff, and curriculum facilities, for the purpose of developing special opportunities for returning veterans and workers. In large cities it may be helpful to set aside a school building to house a special War Service School devoted to high school work for young persons returning from the military service and the war industries. The courses could be accelerated and the calendar fully utilized in order to permit a saving of time. These schools, like others, would grant credit for work completed in the Armed Forces Institute. In smaller cities War Service Schools at the secondary level could be established on a regional basis.

We commend the Regents' plan to educators throughout the Nation.

We commend also the plan whereby men and women in the armed forces may complete academic requirements, while in the armed forces, for graduation from secondary schools. Such educational achievement is possible through work in the service schools, the off-duty educational program, and the educational opportunities of the United States Armed Forces Institute. For men and women who lack a substantial proportion of the requirements for high school graduation, the Army's General Educational Development Tests are helpful in determining the grade level at which service personnel should properly resume their civilian education. The machinery to this end is complete and the procedure is as follows: A complete educational record established while in the service, should be recorded on the official form USAFI No. 47 and returned by the man or woman in the armed forces to the secondary school for evaluation and the award of credit toward graduation. This will facilitate a continuance of education in college of qualified persons. Veterans of World War II who do not file a USAFI credit application form before leaving the service should use a certified copy of their separation record as evidence of in-service training.

6. Importance of Problem of Scientific Training of Men in Armed Forces

The adequate handling of the education of the scientific and technological talent now under arms will be a primary test of the effectiveness of the Government in meeting the whole problem to which we have been asked to direct our attention.

The future scientific and technical leaders in the United States are now largely in military service. Unless exceptional steps are taken to recruit and train talent from the armed service at or before the close of the war, the future will find this country seriously handicapped for scientific and technological leadership. In peace or war, the handicap might prove fatal to our standards of living and to our way of life.

The Educational Pyramid: Studies Concerning Able Students Lost to Higher Education

To be effective, a plan for discovering and developing scientific talent in American youth must be built upon the country's existing educational structure and be consonant with its current operations. Such a plan must recognize the undoubted fact that there is not an unlimited number of individuals of high ability and must ensure that the relatively few with creative capacity in science will be found early and be helped and encouraged to go on through the years of study required to complete professional and research training.

An over-all picture of the child and youth population and of the enrollments in educational institutions is necessary for an understanding of the dimensions of the problem presented by the proposed plan to discover and train young persons of potential scientific ability. Such a picture follows:

The census of 1940 reported the following figures of population under 20 years of age:

	Total	Percent	Male	Percent	Female	Percent
Under 5 years	10,540,524	8.0	5,353,808	8.1	5,186,716	7.9
5 to 9 years	10,684,622	8.1	5,418,823	8.2	5,265,799	8.0
10 to 14 years	11,745,935	8.9	5,952,329	9.0	5,793,606	8.8
15 to 19 years	12,333,513	9.4	6,180,143	9.4	6,153,370	9.4

It will be noted that there are fewer children in the early ages than in later childhood or in the adolescent years. These figures become even more significant in the light of the changes which occurred between 1930 and 1940:

(a) The number of children under 5 years of age decreased, from 1930 to 1940, by some 900,000.

(b) The number of children of ages from 5 to 9 years decreased, from 1930 to 1940, by some 1,900,000.

It is estimated that by 1950 there will be a decrease of some 2 million, and possibly more, in the age group 10 to 19 years. Since the current larger number of babies born during the war will not reach adolescence for another 10 years at least, there will be fewer boys and girls reaching high school and college ages in the next 7 to 10 years.

The following figures from the 1940 census show the age and school attendance of the Nation's 46 million boys and girls and youth:

Age group	Total number	Number attending school	Percent attending
5 years_____	2,142,407	385,160	18.0
6 years_____	2,054,385	1,420,051	69.1
7 to 9 years_____	6,485,830	6,119,026	94.3
10 to 13 years_____	9,340,205	8,915,669	95.5
14 years_____	2,405,730	2,224,670	92.5
15 years_____	2,422,519	2,122,995	87.6
16 to 17 years_____	4,892,170	3,361,206	68.7
18 to 19 years_____	5,018,834	1,449,485	28.9
20 years_____	2,367,042	294,962	12.5
21 to 24 years_____	9,220,793	465,875	5.1

The percentage figures by age groups showing school attendance during 1940 are:

Percent attending school in each age group

Age group:

5 to 6 years_____	43
7 to 13 years_____	95
14 to 15 years_____	90
16 to 17 years_____	68
18 to 20 years_____	23.6
21 to 24 years_____	5.1

It will be observed that the percentage of school attendance rises to age 13 when boys and girls approach the end of the elementary school and junior high school and likewise when employment in many States becomes legal; but that thereafter it declines. From age 17 on, the decline in attendance is rapid, to the 5.1 percent in the years 21 to 24 of college and university attendance.

The following figures show the educational attainments of the population 25 years old and over in the year 1940:

School years completed	Number of persons	Percent
Total [1]_____	73,733,866	100.0
Number school years completed_____	2,799,923	3.8
Grade school:		
1 to 4 years_____	7,304,689	9.9
5 and 6 years_____	8,515,111	11.6
7 and 8 years_____	25,897,953	35.1
High school:		
1 to 3 years_____	11,181,995	15.2
4 years_____	10,551,680	14.3
College:		
1 to 3 years_____	4,075,184	5.5
4 years or more_____	3,407,331	4.6
Median school years completed_____	----------------	8.4

Not including persons for whom school years completed were not reported.

It will be noted that about half of the population 25 years of age and over had completed approximately 8½ grades but that some 13.7 percent had had less than a fifth-grade education.

The foregoing figures of school attendance collected in the 1940 census may be compared with the enrollments by grades in the public schools of the country as tabulated by the United States Office of Education. Again it will be noted that there is a marked decline after the seventh grade and a progressive diminution through the years of high school:

—"Statistics of State School Systems,

	1937–38	1941–42
Elementary school pupils	19,748,174	18,174,668
Kindergarten	607,034	625,783
First grade	3,317,144	2,930,762
Second grade	2,486,550	2,215,100
Third grade	2,444,381	2,175,245
Fourth grade	2,402,617	2,196,732
Fifth grade	2,342,428	2,166,018
Sixth grade	2,252,722	2,124,494
Seventh grade	2,173,173	2,060,752
Eighth grade	1,722,125	1,679,782
Secondary school pupils	6,226,934	6,387,805
First year	1,979,379	1,927,040
Second year	1,669,281	1,705,546
Third year	1,379,398	1,450,788
Fourth Year	1,150,506	1,273,141
Post graduate	48,370	31,090

1939–40 and 1941–42." Biennial Surveys of Education, 1938–40, 1940-42. (Table III, p. 9.)

In the year 1941–42 there was a decrease in high school enrollments of about 189,000, distributed as follows:

First year of high school	84,000
Second year	61,000
Third year	35,000
Fourth year	9,000

Recent reports indicate a larger decrease in high school attendance for the years 1942–43, with indications that some 160,000 boys and some 50,000 girls had left high school. Efforts to reduce the number of students leaving high school and to persuade others to return, have apparently checked this decline in high school enrollments in 1944–45.

There is, as these figures indicate, a progressive reduction in the number of students at each successively higher level of education. Thus, the total student body may be compared to a pyramid with a broad base of elementary pupils sloping upwards to the apex of professional and graduate students.

Various detailed studies of how and when students drop out along the educational sequence have been made which throw light upon the size of and occasion for withdrawals, and the reasons therefor.

Starting with 1,000 pupils enrolled in the fifth grade (figures for earlier grades are confusing because of pupil retardations), the following figures show the extent to which they are reduced in each successive year:

168

Elementary school:
    Fifth grade, 1930-31 1,000
    Sixth grade 943
    Seventh grade 872
    Eighth grade 824
High school:
    First year 770
    Second year 652
    Third year 529
    Fourth year 463
    Graduates, 1938 417
College:
    First year 146
    Graduate, 1942 72

[Statistical Summary of Education, 1939-40 (p. 39)]

In prewar years, of these 72 college graduates, only a few went on to master's degrees and an even smaller number received doctorates.

The foregoing data reveal the gross declines measured in terms of student enrollments. More detailed and individualized studies (cited later) of those who drop out of high school or who fail to go on to college show that there is a significantly large proportion of students of ability, of high level of intelligence, who do not go to college because of lack of funds. In addition it is believed that there are many able, talented, students, whose numbers are difficult to estimate accurately because only a few sample studies have been made, who do not continue their education because schools are inadequate or inaccessible.

Among those who do enter college there is a progressive decrease in each succeeding college year. A study conducted by the United States Office of Education on "College Student Mortality" (Bulletin 1937, No. 11) found that in 1936–37 the percentage of students leaving each year was as follows:

	Percent
In the freshman year	33.8
In the sophomore year	16.7
In the junior year	7.7
In the senior year	3.9

In short, of every 100 students who entered the university in the first year, some 62 withdrew or left before graduation. The figure 62 is, however, a gross figure since it includes 45 students who left to enter other institutions (e.g., students who left after 2 years to enter professional school) or who returned later to the same or went to other institutions of the same level. The reasons for withdrawing or leaving were as follows:

Percentages
    18.4 were dismissed for failure in work.
    12.4 because of financial difficulties.
    12.2 miscellaneous reasons.
    6.1 lack of interest.
    3.4 sickness.
    1.1 disciplinary causes.
    0.8 needed at home.
    0.6 death.
    45.0 unknown.

Those with the lowest academic marks had the highest percentage of withdrawals and those with the highest academic marks had the lowest percentage of withdrawals. But it is to be noted that 12.4 percent, or about one in eight, withdrew because of financial difficulties, indicating that economic need, personally or of the family, was responsible for their leaving college before graduation.

As the foregoing material indicates, students drop out of school in large numbers between the ages of 13 and 14 and likewise during the high school years. Among those who thus drop out there is a significant proportion who have the capacity for further education but who do not continue their schooling. It appears that this premature leaving from high school (and failure to go on to college) arises from—

    Failure to provide educational programs suited to different students who therefore become bored and drop out.

Active recruitment by business and industry of promising youth who for their own good and the national interest should be encouraged to continue their education.

Lack of provision for assisting needy students in high school who must take jobs to help support themselves or their families.

The responsibility for improving these conditions is primarily upon the local community and business interests and the State governments, although it must be recognized that, in some sections of the country, resources are lacking to provide adequate high schools.

A number of recent studies have shown that among high school graduates there are many who have the intelligence and ability for college but who do not go to college for a variety of reasons, chiefly economic and geographic. Excerpts from these studies are given at the end of this section as evidence of the present failure to provide adequately for the continued education of promising American youth.

In the light of the studies made, having regard to the facts of the educational pyramid, it clearly is essential to provide for the early schooling of more able students in order that a large enough group will survive to become a larger quota of high-ability students at the apex of the pyramid. No matter how capable and gifted boys and girls may be, if they do not have opportunities to complete elementary and high school, they cannot go on to college and thence to graduate school for research training.

To increase this small group of exceptionally able men and women it is necessary to enlarge the number of able students who go to college. This involves more and better high schools, with provisions for helping capable students in the high schools (primarily a responsibility of every local community) and opportunities for more capable, promising high school students to go to college. Any other practice constitutes an indefensible and wasteful utilization of higher education and neglect of our human resources.

———

Following are summaries of studies pertinent to our inquiry concerning able students lost to higher education:

The Carnegie Foundation carried out a thorough investigation into the relationships and mutual responsibilities of the high schools and colleges of Pennsylvania. One of the purposes of the study was to answer the question: Who shall go to college?

The procedure of the study involved extensive testing of high school seniors and college students, study of records, and study of progress made in college. Comparisons were then made between college and non-college groups and between various college groups.

The results of this study showed that the group of high school graduates who went to work included many fully as able to obtain high test scores as any pupils who went directly to college. Pennsylvania colleges of arts alone took nearly 4,000 of the high school group tested in 1928. The colleges accepted nearly 1,000 with test scores below the average of the group who did not go to college and they failed to enroll 3,000 with better average scores than the 4,000 they did admit. Although the college group exhibited a test score average superior to that of the non-college group, it did not include the many able and often brilliant high school graduates who could not pay the college bills.

—*The Student and His Knowledge,* by W. S. Learned and Ben D. Wood. Carnegie Foundation for the Advancement of Teaching; Bulletin 29, 1938.

In Minnesota a study was made of students who graduated from high school in 1938 to see what they were doing a year later. About 22,000 young people were included in this study which showed the following:

"What were youth doing a year following graduation? Minnesota high school graduates of June 1938, fall into three broad groups of approximately equal numbers. One-third found full-time employment within a year following graduation. Another third continued their training in either collegiate or preparatory schools. The third group was made up of graduates who secured part-time employment only, of the unemployed, and of those graduates for whom principals were unable to supply information.

"Of the 22,306 young people who finished high school in June 1938, 35 percent were employed full time and 7 percent had secured part-time employment in April 1939; 12 percent were unemployed; 23 percent were enrolled in colleges or universities, and an additional 12 percent were receiving training in other kinds of schools—trade schools, commercial colleges, schools of nursing, high schools as postgraduates. High school principals were unable to report the whereabouts of 11 percent.

"Was there a relationship between scholastic achievement in high school and employment or further training for Minnesota high school graduates? When the relationship of scholastic success in high school to the post-high school status of the graduates of June 1938 was studied, these three trends were found: As one goes down the ability scale (1) the percentage of graduates employed increased, (2) the percentage of unemployed graduates also increased, and (3) the percentage of graduates who continued their training beyond high school decreased. When, however, only those graduates who presumably were in the labor market (not continuing their education) were considered, high school success bore little relation to employment and unemployment.

"Many able graduates, however, were not attending college. Considerably less than half of the high school graduates who ranked in the upper 30 percent of their high school classes were enrolled in college. More than 15 percent of these able graduates who did not continue their training were unemployed. High marks in school are doubtless desirable, but they were not the open sesame to college halls or employment for those graduates" (p. 35).

For every (high school) graduate who ranked in the upper 10 percent of his high school class and entered college, another graduate who also ranked in the upper 10 percent did not enter college.

For every graduate who ranked in the upper 30 percent of his class and entered college, two graduates who ranked in the upper 30 percent did not enter college.

"Was there a relationship between socio-economic status as indicated by the fathers' occupations and the sta-

tus of Minnesota young people a year following their graduation? From the professional end of the occupational scale to the unskilled labor end, (1) employment increased, (2) unemployment increased, and (3) the proportions of graduates continuing their training decreased. From this study of the Minnesota high school graduate of June 1938, it would seem that the probability of college attendance for a graduate who is the son or daughter of a father employed in a professional or managerial group is several times greater than that for the son or daughter of a farmer or of an unskilled laborer. Among the June 1938 graduates, children of the unemployed were themselves unemployed in greater proportion than children of fathers at work" (p. 36).

Many able high school graduates were not enrolled for further education. "It is no longer safe to assume—if it ever was—that the most intelligent high school graduates go to college. It is of fundamental importance for all the people of the State to know how generally young people who would make the best teachers, lawyers, accountants, doctors, engineers, and statesmen are able to attend colleges and universities. It has been assumed traditionally that the most capable high schol graduates go to college. It is suggested by this study, however, that geography and the economic resources of the family are perhaps as closely related to college attendance as is intellectual fitness" (p. 39).

—"What Happens to High School Graduates?" by G. Lester Anderson and T. J. Berning. Studies in Higher Education. Biennial Report of the Committee on Educational Research

1938-40. University of Minnesota, 1941.

———

"It is possible to investigate the availability of educational opportunity * * * in various parts of the country. For example a study of youth in Pennsylvania was conducted about a decade ago by the State Department of Public Instruction and the American Youth Commission. The socio-economic status and educational history were ascertained for a group of 910 pupils with intelligence quotients of 110 or above. It is generally assumed that pupils with intelligence quotients above 110 are good college material. This group of superior pupils was divided into two subgroups on the basis of socio-economic status. Of the upper socio-economic group, 93 percent graduated from high school and 57 percent attended college. Of the lower socio-economic group, 72 percent graduated from high school and 13 percent attended college. Further study of the data in Table II will show even more clearly that the group with below-average socio-economic status had relatively less educational opportunity than the group with above-average socio-economic status, although both groups were about equal in intellectual ability" (p. 51).

"A similar conclusion must be drawn from a study made by Helen B. Goetsch on 1,023 able students who graduated from Milwaukee high schools in 1937 and 1938. These students all had I. Q.'s of 117 or above. The income of their parents is directly related to college attendance, as is shown in Table III. The higher the parents' income, the greater is the proportion who went to college" (p. 52).

172

"Table II: Relation of Intelligence to Educational Opportunity

[Record of students with intelligence quotients of 110 or above]

Educational advance	Socio-economic status above average		Socio-economic status below average		Total group	
	Number	*Percent*	*Number*	*Percent*	*Number*	*Percent*
Dropped school at eighth grade or below	4	0.07	27	7.9	31	3.4
Completed ninth, tenth, or eleventh grade but did not graduate from high school	36	6.2	69	20.2	105	11.6
Graduated from high school but did not attend college	206	36.3	202	59.0	408	44.8
Attended college	322	56.8	44	12.9	366	40.2
Total	568	100.0	342	100.0	910	100.0

"Table III: Relation of Parental Income to Full-Time College Attendance of Superior Milwaukee High School Graduates

Parental income:	*Percent in college full time*
$8,000+	100.0
$5,000–$7,999	92.0
$3,000–$4,999	72.9
$2,000–$2,999	44.4
$1,500–$1,999	28.9
$1,000–$1,499	25.5
$500– $999	26.8
Under $500	20.4

"We see what actually happens if we consider the 191 students who were graduates of the Old City High School over a 5-year period. This number includes all the white high school graduates except those who attended private schools. Table IV shows what happened to these people after graduation and what the social make-up of the group was" (p. 59).

"Table IV: College Attendance of High School Graduates in Old City

Class	Number	Percent of total by class	Number attending college	Percent of each social class attending college	Percent by social class of all who attend college
Upper	14	7	10	72	14
Upper middle	54	28	37	69	51
Middle	31	16	18	58	25
Lower middle	43	23	7	16	10
Lower	19	10	0	0	0
Unknown	30	16	0	0	0
Total	191	100	72	---------	100

"The Hometown school has a fine building and an undifferentiated curriculum so that the same high school education is available to all the children, whether they have college ambitions or not. In Hometown, 80 percent of the boys and girls of high school age attend high school. Why do they go? What do they and their parents expect from a high school education?

"First of all, no upper-upper class family has children in high school. The lower-uppers and upper-middles account for about the same proportions of pupils as one would expect from their proportions in the total population. The lower-middles contribute less than one would expect and the upper-lower and lower-lower contribute more, probably because the lower-class people have larger families and therefore, more prospective pupils.

"Of all high school students classified as lower-upper or upper-middle, 88 percent will go on to college while only 12 percent of those in the three bottom classes expect to go to college. Of the total high school pupils, 20 percent are preparing to go to college and 80 percent were definitely not going to college" (p. 66).

"The generalization that different curricula and types of institutions are adapted to different statuses is illustrated by Goetsch's study. She found that the hierarchy of family income was reflected in a hierarchy of courses pursued by students in higher institutions, as shown in Table VI" (p. 72).

"Table VI: Parental Income and College Courses

Curriculum:	Median parental income
Law	$2,118
Medicine and Dentistry	2,112
Liberal Arts	2,068
Journalism	1,907
Engineering	1,884
Teaching	1,570
Commercial	1,543
Nursing	1,368
Industrial Trades	1,104

—Who Shall Be Educated: The Challenge of Unequal Opportunities, by W. Lloyd Warner, Robert J. Havighurst, Martin B. Loeb. Harper & Bros., New York City, 1944.

"The findings of this study, in harmony with the findings of other studies, show that approximately as many of the ablest high school graduates are out of college as are in college.

"On the basis of the sample (of 1,754 cases), the upper quarter of the State's 16,000 high school graduates would contain a minimum of 4,000 of the ablest individuals, the type of students who really do well in college. Forty-nine percent of 4,000 is 1,960 individuals with high potential college ability, who for some reason or reasons, did not enroll in college. From the point of view of the colleges, as well as of the individuals and of society, the loss in human resources indicated in these data is highly significant.

"Table 8 shows that for every four

able boys in the upper quarter there were six able girls. Table 11 shows that the ratio of able boys to able girls in the upper quarter enrolled in college was 6 to 4.5. Thus, it is clear that the greatest social and personal loss of human resources comes in the ranks of able girls in the upper quarter" (pp. 37-38).

—"The Utilization of Potential College Ability Found in June 1940, Graduates of Kentucky High Schools," by Horace Leonard Davis. Bulletin of the Bureau of School Service, College of Education, University of Kentucky. Vol. XV. No. 1. Sept. 1942.

"Location of brightest seniors. When we determine which economic group furnished the largest percentage of seniors possessing the higher grades of intelligence we secure different results. All economic groups except the highest salaried group are represented in the highest one percentile class. Table XLII shows the percentage of students belonging to each economic group whose mental test score gave them a rating of A+

or A, the highest grades made on the tests, also the percentage making a mental rating of E— or F, the lowest grades of intelligence possessed by our total or standard group. Groups 2, 3, and 4, where the income varied from $1,000 to $4,500, have the largest percentage of seniors rated A+ and groups 1 and 5 the smallest. Groups 3 and 4, are superior to group 2 in the percentage of students rated A+ or A.

"From a study of our distribution tables it appears that neither group 1 nor group 5 contain students who score above 180 points in the tests. But seniors possessing this grade of ability were found in each of the other economic groups. The brightest students belong to group 4, the annual income of whose parents ranged from $1,000 to $2,000. Eight students belonging to the group, 6 boys and 2 girls, made scores over 185. And 5 students in our lowest economic group (annual income $500 to $1,000) made scores ranging from 175 to 180, while there was but a single student in our highest salaried group who made a score above 175 points" (p. 213).

"Percent of Students in Each Economic Group Possessing Highest or Lowest Grades of Ability

Economic groups compared	1 Salary $4,500–$12,000	2 Salary $3,000–$4,500	3 Salary $2,000–$3,000	4 Salary $1,000–$2,000	5 Salary $500–$1,000
Percent rated:					
A+	1.89	3.01	2.48	2.49	0.82
A+ or A	7.56	7.53	10.07	8.24	5.68
F	1.08	1.50	.55	.81	1.92
E— or F	5.13	6.52	4.69	5.03	8.99
Total cases	370	199	724	1,964	1,089

"If we count all students whose test scores gave them a mental rating of A+, A or B, these various economic

groups arrange themselves . . . : Group 3 (income $2,000 to $3,000) comes first; group 2 (annual income

175

of $3,000 to $4,500) comes second; group 4 (annual income of $1,000 to $2,000) ranks third; while the highest and lowest salaried groups come last" (p. 214).

"If we compare the records made on our tests by the group of seniors representing the richest and the poorest homes, we find that there are proportionally more children possessing the highest grades of mental ability among the poorest class than among the wealthiest class, and more individuals with high average grades of intelligence among the wealthier than among the poorer group. The wealthiest group ranks high on central tendency. The poorest salaried group ranks low on central tendency and also has a larger percentage of individuals possessing the lower grades of mental ability. But there are individuals in this class who obtain the highest intelligence rating made by high school seniors" (p. 216).

"*Brightest seniors not going to college.* It is still more significant that so many of this most superior group of high school seniors will not attend college, while those with the most inferior grades of intelligence are planning to attend, in ever increasing numbers. Twenty-five percent of the brightest seniors found in the entire State said they were not planning to attend college at all, while 65 to 70 percent of the dullest seniors had definitely decided to go to college, most of them having already selected the college they expected to attend" (p. 298).

—*The Intelligence of High School Seniors,* by William F. Book. The Macmillan Company, New York, 1928.

Data Concerning Training of Personnel for Science and Technology

The relatively small number of able students who graduate from college must be shared by the various professional schools and by the graduate schools which train for research in the natural sciences, the social sciences, and the arts and humanities.

From compilations made by the American Association of Collegiate Registrars, the following figures show the distribution of undergraduate degrees in 1941 among broad academic and professional fields:

Social sciences _____ 17,947
Social science and allied fields of law, business administration, education, divinity, library training, journalism, etc._____ 70,829
Mathematics and physical sciences 6,440

Mathematics, physical science and allied fields of engineering, mining, chemistry, etc._____ 25,044
Biological sciences _____ 5,812
Biological sciences and allied fields of medicine, nursing, dentistry, pharmacy, agriculture, forestry, home economics, etc._____ 28,490

In the fields of research these fall into broad groups, as follows:

Social science _____ 17,947
Natural science _____ 16,050
Natural science and technology_____ 53,534

A compilation of Ph.D. degrees in the 6 years before the war shows the following distribution among the physical, earth, biological and medical sciences, psychology, public health, and anthropology:

Subject	1935	1936	1937	1938	1939	1940	Average
Astronomy	11	5	9	12	5	6	8
Chemistry	470	482	497	426	482	527	479
Engineering	63	48	70	59	44	77	60
Mathematics	77	84	76	62	91	103	82
Metallurgy	11	16	7	7	9	11	10
Physics	150	147	158	165	148	191	160
							799
Geology	62	64	42	58	49	55	55
Meteorology	1	0	1	4	2	0	1
Mineralogy	1	5	3	5	1	4	3
Seismology	2	----	----	----	----	2	1
							60*
Paleontology	12	10	8	9	13	11	11
Biochemistry	----	----	----	101	127	130	(120)
Agriculture	77	53	48	37	40	58	52
Anatomy	25	15	14	20	17	21	19

Subject	1935	1936	1937	1938	1939	1940	Average
Bacteriology and Microbiology	38	41	46	40	56	59	47
Botany	110	108	88	106	108	112	105
Entomology	34	30	51	33	47	48	41
Genetics	10	21	13	31	32	26	22
Horticulture	24	14	21	16	11	20	18
Physiology	76	83	103	66	59	70	76
Zoology	113	132	98	102	102	112	110
							621
Medicine and Surgery	14	12	1	7	9	10	9
Pharmacology	10	18	14	19	23	23	18
Psychology	101	118	112	108	123	120	114
Public Health	4	13	9	15	8	15	11
Anthropology	13	20	15	18	11	26	17
							169
Total							1,649

It will be noted that the physical sciences provided about 800 or almost half of the total number of doctor's degrees in science, of which, in turn, about one-half were in chemistry. The earth sciences provided an additional 60 degrees.

After the physical sciences the largest number of degrees were in the life sciences, with about 800 degrees distributed among the several divisions as follows:

Medical sciences	300
Biology	307
Agriculture	52
Psychology	114
Anthropology	11
	784

The National Roster made an inventory of graduate students in non-professional and non-vocational schools and departments, as of December 1942, which showed the number then enrolled in the graduate schools, divided into disciplines as follows:

Physical sciences	5,698

Chemistry	3,045
Geology	182
Mathematics	545

Meteorology	918
Physics (electronic)	227
Physics (non-electronic)	680
Other physical sciences	101
Biology	1,120
Social sciences	3,857

Economics	1,034
Geography	79
History	812
Psychology	543
Other social studies	1,389
Languages, literature, fine arts, and music	2,157
Other major fields	486
	13,318

It will be noted that there were about 5,700 graduate students in physical science and some 1,100 in biology at the time of this report.

In terms of the educational pyramid, the total number of graduate students shown above in all branches of non-professional and non-vocational graduate study form but a small proportion of the total college enrollments of approximately 1,400,-000 in 1939–40. Moreover, the 1,649 who received Ph.D. degrees in the sciences listed above were about one-half of the total number of all Ph.D. degrees (about 3,300 in 1939–40).

Special studies have been made by the Institute of Physics on the effect of war upon the training of research personnel in the graduate schools. These indicate that there is a cumulative deficit in the number of students receiving Ph.D. degrees, in the several physical sciences and engineering, which will continue and grow until several years after the war:

	Accumulated deficit, 1941 through 1944	Estimated deficit, 1945	Total 1941 through 1945	Probable deficits, 1946 through 1955	Total probable deficit due to war, 1941 through 1955
Chemistry	240	550	790	4,460	5,250
Engineering	148	82	230	730	960
Geology	63	50	113	317	430
Mathematics	161	100	261	939	1,200
Physics	251	160	411	1,589	2,000
Psychology	96	84	180	550	730
Biological sciences	665	725	1,390	4,910	6,300
Totals	1,624	1,751	3,375	13,495	16,870

Proposals for enlarging the number of students entering the graduate schools to be trained for research must be considered in relation to the probable demand for trained research workers as expressed in available jobs. Thus, necessary as it is to enlarge the number of graduate students in order to produce the relatively few research students of exceptional ability, the danger of an oversupply of trained research personnel must be kept constantly in mind. The unhappy plight of scholars in Europe after the last war when there was a surplus should not be forgotten.

Likewise proposals for recruiting more college students into the physical and biological sciences and enlisting more graduate students for training in research in the physical and biological sciences should be viewed in the light of the over-all needs of the country and of the requirements in other fields of research and in the several professions. If too many of the limited number of high quality students are absorbed by fields of scientific research, research in the social sciences and in the arts and humanities may be jeopardized with probably unfavorable reactions upon scientific research.

Appendix C

Suggested Administrative Organization, Bases of Selection, Schedule and Procedures

I. *Administrative Organization*

Proposals for a National Scientific Research Foundation are under discussion, such a foundation to be charged with responsibility for the administration of the several national scientific programs being recommended by the committees advising Dr. Bush. Among the suggested powers of such a foundation is power to make contracts with other agencies for the performance of functions within the scope of the foundation. It would be our recommendation that the foundation, or any similar agency which may be established, should make arrangements for choosing Scholars and Fellows under the proposed plans through the National Academy of Sciences, if that organization be willing to accept the responsibility. The National Academy, as a representative body of the scientists of the country, would be the logical organization to sponsor this program; and association with the National Academy would give the program respect and prestige.

It is further recommended that:

(*a*) The President of the National Academy of Sciences, with the advice and consent of the Administrative Committee of the National Research Council, periodically should appoint a National Science Reserve Advisory Committee which would advise him regarding methods of selection and the arrangements for the entire program within the provisions of the legislation.

(*b*) The President of the National Academy of Sciences with the advice of the advisory committee should appoint a Director of the National Science Reserve Program, who would devote his full time to the work. Subject to the general supervision of the administrative committee of the National Research Council, the Director would establish a national office, select the administrative and research staff, develop detailed plans and administer the program. The Director would appoint a technical advisory committee, or committees for the development of tests of scientific promise, of the inventory of activities and interests and of the recommendation blank and rating scale.

(*c*) The President of the National Academy of Sciences with the advice of the advisory committee would appoint State committees of selection, to consist of five members, *to wit*: three scientists, one of whom should serve as chairman; one member of a college or university faculty, trained and experienced in the field of selection and guidance; and one representative of secondary education, usually a school principal or one of the high school supervisors in the State department of education. At least one of the scientists, it is suggested, should be from agriculture, or from industry, within the State. The State committees of selection, under our plan, would have the responsibility of making the final appointments to the limit of the State quotas. These

180

committees would make their selections on the basis of the dossiers of the candidates which would be supplied to them by the national office, plus such other material as the State committees may decide to gather and use. The national office, on the basis of the tests of scientific promise and the applicants' school records, would select twice the State's quota and report the names and records of such candidates to the State committees. The State committees would make the final selections from among these candidates. No candidates who do not attain a certain minimum national standard should be certified to the State committees.

In the selection of the general advisory committee, and also in choosing the membership of the technical committee or committees, the persons and organizations that have had the greatest experience in constructing, administering and interpreting the particular kind of measuring instruments to be used in this program should be consulted. Such organizations include the committee on Measurement and Guidance of the American Council on Education, the Cooperative Test Service, the College Entrance Examination Board, the Graduate Record Examination Office of the Carnegie Foundation, the Measurement and Guidance Project in Engineering Education, the Examination Staff of the Armed Forces Institute, and the University of Iowa Examination project. Directors, and in some instances, other staff members of these agencies are among those who have had the greatest amount of experience in preparing measurement instruments of the type necessary. To make certain that the selection instruments are as adequate as they can be made, it will be essential to draw upon the combined experience and technical knowledge and judgment of these persons and agencies. The whole job must be done at the highest possible level of professional competence.

II. Bases of Selection[1]

It is proposed that there be four principal sources of information and judgment upon which final selection of the Scholars should be based; but that only the first two of these be used in the preliminary screening:

(1) Score on test of scientific promise.

(2) School record, especially rank-in-class.

(3) Candidate's application including an inventory of activities and interests.

(4) Recommendation of principals and teachers regarding candidate's ability and personal qualities.

It is strongly recommmended that these tests and examinations be undertaken on an experimental basis and be continually revised and improved in the light of actual experience and of the performance of students selected. The first few years of the program especially should be considered experimental so that new and promising methods can be tried out, particularly for the discovery of the candidates' interests and personality characteristics, including evidence of some concern for social understanding and responsibility.

For the present, and subject to change in the light of experience and

[1] This section applies especially to the discovery of talented youth who are attending high school. The committee, as indicated in the body of its report, recognizes that there is also the problem of finding, and giving opportunity to, talented youth who are outside high schools and that, for them, variations from standard procedures will be required.

research upon the validity of indices of prediction, it is recommended that:

The test score and rank-in-class in school should be combined into one index of academic promise which should be used as the basis of screening. For each State a critical score on the index should be set at a point which would yield twice the State quota, provided that the State critical score were above the national minimum score.

For the candidates above the State critical score, additional information should be collected so that in the final selection it will be possible to take into account certain important qualities such as originality, creative ability, motivation, emotional stability, and qualities of leadership.

For the convenience of the State committees of selection, a summary sheet would be prepared giving the essential data from the four sources of information indicated above and this summary sheet would be attached to the front of each candidate's dossier when it is sent to the State committee.

The State committees should be provided with directions to assist them in interpreting the various items of information about each candidate. The State committees, however, should be entirely free to use and evaluate the information in accordance with their best judgments and should be encouraged to collect additional information, such as interview reports, concerning the applicants, to provide the broadest possible bases for the process of selection.

1. Test of Scientific Promise

(a) *Length.*—The test should be of sufficient length for efficient selection, perhaps of 5 or 6 hours duration. It should not be a speed test.

(b) *Type of questions.*—The test should be of the objective or controlled-answer type. The unreliability of free answer questions as well as the limitation in sampling imposed by such questions restricts their usefulness for the present purpose.

(c) *Content.*—There should be several sections in the test. The materials throughout should be such as to involve a complex of aptitude and achievement as the most satisfactory measure. The subject matter should be related particularly to scientific ability instead of to general academic promise.

(d) *Level of difficulty.*—It is important that the test be of maximum selectivity at and above the critical score. Studies of the results of the best mathematics and science tests now in use indicate that students who make very high scores on such tests can be expected to succeed in scientific courses during the first year of college with a high degree of certainty.

(e) *Preparation of the tests.*—(1) The test should be prepared after consideration of the specifications recommended by the advisory committee which would include both scientists and testing specialists. (2) The test material should be pretested on a suitable population and the final test made up of the most successful items.

2. School Record

The school record is important because it is a measure not only of ability but of the application of that ability in academic work over a period of several years. It has been found to be as useful in predicting college success as an aptitude test and when combined with the test score, the combination provides an index that is superior to either item

182

used alone. The school record not only adds the element of industriousness but rank-in-class and test score are what might be described as automatically compensatory indices. Aptitude test scores are not entirely independent of the home and school background. Although aptitude more than achievement is measured, no aptitude tests are "pure" and uninfluenced by previous training. Consequently individuals attending "good" schools are likely to be somewhat overrated by their test scores. For such students, their rank-in-class score is likely to be an underestimate of their achievement. Boys and girls from inferior schools on the other hand are likely to be underrated by their test scores and overrated by their rank-in-class. The two indices combined, therefore, provide a fairer basis for screening than either alone.

The most reliable single measure of school success is rank-in-class. Marking systems and standards differ from school to school but the significance of relative standing in class remains fairly constant. Pragmatically it has been found to be the best index of school achievement.

For all candidates who are above the State critical score on the screening index, complete transcripts of high school records should be obtained along with other information to be used by the State committees in making the final selections.

3. Candidate's Application, Including an Inventory of Activities and Interests

The application blank which candidates who pass the screening test will be required to fill out will contain (a) the questions concerning age, family, schools attended, etc., such as are usually asked on a college admission blank; (b) an extensive inventory of activities and interests specifically prepared for this purpose; (c) a statement of the extracurricular scientific activities of the student during the preceding 2 years.

Inventories of activities and interests, while not yet in as high a stage of development as intelligence tests, are valuable in indicating personal and intellectual qualities not measured by tests. Indications of such qualities as the individual's ability to get along with others, his scientific interests and motivation, his emotional stability may be obtained from the inventory. These indications should be checked against the ratings of the principal and teachers on those same qualities.

An advisory committee composed of men who have specialized in this field of measurement should be appointed to draw up the specifications for the inventory of activities and interests to be developed by the staff of the national office.

4. Recommendation of Principal and Teachers Regarding Candidate's Ability and Personal Qualities

Although the principal and teachers are not entirely impartial persons from whom to receive recommendations, they have a better comparative basis for making judgments and are more likely to make fair and frank estimates than others whose opinion of the candidates might be requested.

The recommendation blank should be in two parts. The first part would contain a large number of multiple choice questions and rating scales, the answer to some of which would relate to specific observations on points of fact, while others would involve judgment of less tangible qualities. Each teacher who has had the candidate in class or in extra-curricular activities would indicate his or her answer by initialling the blank. The principal would finally, with an X, indicate the consensus. The second part of the blank would call for statements regarding a few particularly important qualities, such as concrete evidence of originality or creative ability in the field of science.

Part one would yield indications on the same qualities as would be indicated by the candidate's inventory of activities and interests, so that the two sets of scores could be considered together and serve as a check on each other.

The advisory committee charged with the responsibility of preparing the specifications for the inventory of activities and interests should also prepare the specifications for the recommendation blank.

III. Schedule and Procedures (Tentative)

June 1–Oct. 1	Preliminary publicity through newspapers, magazines, and radio.
Oct. 1	Announcement by letter to State departments of education, superintendents of schools, and principals, giving the detailed plans of the competitions. School principals would be sent a return postal card on which to indicate the number of students in the school who would take the screening test and the names of the teachers who would administer it.
Oct. 20	Return postal cards due at national office.
Oct. 20–Nov. 15	Screening tests and directions for administration shipped to schools.
Dec. 1	Screening tests administered. A detachable portion of the answer sheet containing the same serial number as the answer sheet would be turned over to the principal after the candidate has filled in his name and the name of the school so that the principal can enter the candidate's rank-in-class. (Complete directions for the principal will be printed on the form.)
Dec. 5–Jan. 10	Scoring of tests and calculation of composite index for screening.
Jan. 10	Notice to candidates of success on screening test. Letter to principals of schools having successful candidates with application and recommendation blanks enclosed. Full directions will be given the principal for the administration of the inventory of activities and interests and for the filling out of the recommendation blank.
Feb. 10	Application and recommendation blanks due at the national office.
Feb. 10–Mar. 10	Candidates' dossiers put in order and shipped to State committees of selection.
Mar. 20–Apr. 10	State committees go over applications, gather additional information (if they wish), and make selections.
Apr. 10	List of men and women selected for scholarships sent to national office.
Apr. 15	State committees notify candidates of selection for scholarships.
Apr. 16	Public announcement of selection of scholars.
May 1	Successful candidates must notify State committee of acceptance, of the university or college they wish to attend and of the science course they wish to pursue.

184

May 15	State committee notifies appointees of approval of college and course of study and corresponds with those where approval is withheld.
May 30	Final revised list of appointees with name of college they will attend and course of study they will pursue sent by the State committee to national office.
	(From this point on, appointees deal directly with national office.)

IV. Minimum Annual Cost of Selection (Estimated)

Screening: 200,000 candidates at $1	$200,000
Final selection: 12,000 candidates:	
National office at $5	60,000
State committees at $2	24,000
Research and experimentation (average)	50,000
	334,000

It is of the utmost importance that adequate funds be allocated for research on the methods of selection. Although present knowledge makes it possible to do an effective job in selecting youth of scientific promise, work in this field is still in the early stage of development. A strong research program would certainly lead to improvement in the selection of future scientists and in view of the suggested size of the program would be a long-run economy.

Appendix 5

Report of the Committee on Publication of Scientific Information

Table of Contents

LETTER OF TRANSMITTAL

DR. VANNEVAR BUSH, *Director,*
Office of Scientific Research and Development,
1530 P Street, NW., Washington, D. C.

MY DEAR DR. BUSH:

It is my pleasure to submit herewith the report of the Committee appointed to assist you in answering the first question in President Roosevelt's letter to you of November 17, 1944, which was expressed as follows:

> "*First*: What can be done, consistent with military security, and with the prior approval of the military authorities, to make known to the world as soon as possible the contributions which have been made during our war effort to scientific knowledge?
>
> "The diffusion of such knowledge should help us to stimulate new enterprises, provide jobs for our returning servicemen and other workers, and make possible great strides for the improvement of the national well-being."

In preparing the report the members of the Committee had the benefit of discussions with a number of persons concerned with the publication of scientific information. There has been general agreement that one of the primary problems in the field of publication is the establishment of an agency which, as a general principle, will permit the release of scientific information as soon as it can no longer be used against us in the present war and on terms which will be fair to all concerned. In particular, speed of release should be accompanied by a mechanism which will lift the restrictions on publication in a particular field uniformly for all workers in that field, regardless of the particular agency of the Government for which the work might originally have been done. The Committee feels strongly that this mechanism should be established without any unnecessary delay.

Sincerely yours,

IRVIN STEWART,
Chairman, Committee on Publication
of Scientific Information

MEMBERS OF THE COMMITTEE

Dr. Irvin Stewart, chairman; director, committee on scientific aids to learning, National Research Council; executive secretary, Office of Scientific Research and Development.

Dr. J. P. Baxter III, president, Williams College.

Dr. Karl T. Compton, president, Massachusetts Institute of Technology.

Dr. James B. Conant, president, Harvard University.

Dr. A. N. Richards, vice president in charge of medical affairs, University of Pennsylvania.

Dr. M. A. Tuve, physicist, department of terrestrial magnetism, Carnegie Institution of Washington.

Mr. Carroll L. Wilson, executive assistant to the director, Office of Scientific Research and Development.

Mr. Cleveland Norcross, secretary, executive assistant to the executive secretary, Office of Scientific Research and Development.

REPORT

The following report is submitted in answer to your request for advice with respect to the first point in President Roosevelt's letter to you of November 17, 1944.

1. *Need for Lifting Restrictions*

The frontiers of science must be thrown open so that all who have the ability to explore may advance from the farthest position which anyone has attained. During the war we have been living to a considerable extent on our scientific capital, as scientists who would normally be extending the frontiers of knowledge have instead devoted their efforts to the application of our scientific knowledge to the development of new and better equipment, processes, and materials for war purposes. A large part of such new scientific discoveries as have been made, together with the great amount of information on the techniques of application, are now classified as confidential or secret. The restrictions incident to war have prevented the wide spread of the kind of information upon which American science, education, and industry normally build. Scientists engaged on war projects have acquired new knowledge in specific fields, but they have not been given access to similar acquisitions by their colleagues in other fields. Thus, while there is a fund of new knowledge scattered among a large number of individual scientists, no one of them has access to all of it; and the broad base of scientific knowledge available to all scientists has not been correspondingly extended. This situation should be speedily corrected.

During the first year of the existence of the Office of Scientific Research and Development a decision was made by the Secretaries of War and Navy that in the fields of medical research, publication of new knowledge should be withheld only if that knowledge gave promise of conferring military advantage. Hence it has been possible to publish most of the newly developed knowledge in the medical field. Several hundred articles have already been published in the professional journals and others are in the process of publication. The amount of classified medical material has been held to a minimum. It has been confined largely to limited subjects of immediate battle front importance and to information which might be related to strategy. Even these limited restrictions should be lifted as soon as military conditions permit.

Not all of our troops can be returned immediately upon the cessation of hostilities. Many men must remain overseas, some in armies of occupation, others awaiting the provision of facilities for their return. Educational facilities must be provided for them during this period. Very recent techniques developed in our laboratories in connection with the prosecution of war developments can and should be made available in the Army universities overseas to qualified men in order that they may thereby be enabled to return to this country with as modern and advanced approach to some of the subjects of moment as they would have had if they had remained here during the war, or if they had been selected for early return and re-entry into universities in this country. To accomplish this not only must the information be available in printed form, but men familiar with latest developments should be chosen as instructors in the Army universities.

The returning soldier who wants to pick up his interrupted plans for a career as a scientist or engineer deserves access to the very latest developments and techniques. It will be a tragedy for him and for the country if he is trained in the light of the knowledge of 1940 rather than 1945. Because of the war we have lost several classes of scientists and engineers, both undergraduate and graduate. The gap can never be entirely filled, and it can be successfully narrowed only if the classes graduating in the immediate postwar years can be trained in advanced developments and techniques. We must overcome, not aggravate, the effects upon science and upon the country as a whole of the wartime loss of several classes of scientists.

These considerations merit emphasis in addition to those mentioned in the President's letter of November 17.

2. *Release from Military Classification*

The first, and most important, step is to obtain the release of scientific material from its military classification as soon as conditions permit. Basically there is no reason to believe that scientists of other countries will not in time re-discover everything we now know. A sounder foundation for our national security rests in a broad dissemination of scientific knowledge upon which further advances can be more readily made than in a policy of restriction which would impede our further advances in the hope that our potential enemies will not catch up with us. The Committee believes that, with few exceptions, our national interests require the release of most of our war-acquired scientific information as soon as it is evident that our enemies will not be able to turn that information against us in the present war. It further believes that most of this information can be released without disclosing its embodiments in actual military material and devices.

Research has gone forward under many auspices, the Army, the Navy, the National Advisory Committee for Aeronautics, the Office of Scientific Research and Development, various other Government departments and many industrial establishments and academic institutions. In many cases there have doubtless been independent discoveries of the same truth in different places. To permit the release of information from one place and restrict it from another would not only be unfair but would impair the morale and efficiency of scientists who have readily subscribed to the policy of restriction dictated by war needs.

190

The agency charged with the duty of recommending release of information from military classification should be a continuing one well grounded in science and technology, which can couple advice to the military with an ability to obtain prompt decisions. With that in mind you have recently proposed the establishment within the National Academy of Sciences of a board to control the release and promote publication of certain scientific information. Its standing at the apex of the scientific world together with its contributions to the present war qualify the Academy in a unique manner to perform this service. The proposed board with its joint Army, Navy, and civilian membership should be able to act promptly and intelligently, with full appreciation of both military and civilian implications of its decisions. It should provide the speed which is essential if delay is not to nullify a large part of the benefit sought by the release of newly discovered scientific information. Obviously the board should be adequately manned to act promptly.

3. *Agreement with Our Allies on Release of Information*

Some of the information which should be released is possessed jointly by our allies and ourselves. Release in this country should be coordinated with release in other countries where the restriction has been jointly imposed in both. A central agency such as the proposed board should be able to handle this normally time-consuming but important matter with a minimum loss of time and danger of international friction.

4. *Stimulation of Publication*

It is obvious that the contributions to scientific knowledge "made during the war effort" fall into many categories. Much of this information is now being made public through various media as, for example, most of the results of medical research. This report is directed to those contributions to scientific knowledge which are prevented from being "made known to the world" because of Government restrictions. Most of this information resulted from work in which some Government agency was interested and is now under security classification. The two chief obstacles to prompt publication are: (1) security regulations; (2) the policy of cognizant agencies in releasing investigators to publish freely. A courageous policy on the part of administrative officers of Government agencies in assisting and stimulating prompt publication by Government scientists as well as private contractors and their employees as soon as security regulations are relaxed will cover point two. The first point, we believe, can be covered by the creation of the board to control the release and promote the publication of certain scientific information.

The object is to get the scientific results of war research written by outstanding experts, completely available, especially to young scientists, at as low a cost to them as is consistent with doing the job well.

In connection with scientific war research being performed under contracts of governmental agencies, which has necessitated bringing together large groups of scientists, the most advantageous time for preparation of manuscripts may well be during the final months of the contract, while the scientific staffs are still assembled and in possession of all records, but after the pressure for production of war results has begun to relax.

Obviously not all reports will merit publication and distribution. Where Government-financed research is involved, the contracting agency must make the decision. In every case, however, this decision should be made upon the basis of the public interest in the dissemination of the information, not upon the presence or absence of funds to defray the cost of publication and distribution of the report.

The publication plans of the Office of Scientific Research and Development are being made in accordance with the principles of the preceding paragraphs. The effectiveness of these plans, as well as the publication of other scientific information developed in connection with war research, will depend largely upon the speed with which the proposed Academy board is established and the effectiveness with which it functions after its establishment. The impetus which has produced remarkable results in the laboratory and in the field will be lost if publication is unduly delayed.

5. *Recommendations*

In specific answer to the first point in the President's letter, therefore, your Committee recommends the following:

1. The prompt establishment and adequate staffing within the National Academy of Sciences of the proposed board to control the release and promote publication of certain scientific information. This is essential.

2. The adoption by that board of a liberal policy generally permitting the release of scientific information as soon as it is apparent that such information cannot be turned against us in the present war.

3. The encouragement of scientists to publish the results of their investigation in "open" fields covered by releases by the board.

4. The stimulation and assistance of investigators to prompt publication by administrative officers of cognizant Government agencies.

5. The provision of adequate financing for the publication and distribution of the reports mentioned in the preceding paragraph.

INDEX

INDEX

Abstracting services. See Libraries; Reference aids.

Adams, John Quincy, support of ideas for Government-sponsored academy, 84.

Advisory Board. See Science Advisory Board.

Advisory committees, recommended for each Government bureau engaged in scientific work, 105-106.

Aerodynamics, future progress in, to be striking, 78.

Aeronautical and Space Sciences, Senate Committee on, viii.

Aeronautics. See National Advisory Committee for Aeronautics; National Aeronautics and Space Council; Science and Astronautics, House Committee on.

AFL-CIO, conference on "Labor and Science in a Changing World," xi.

Agassiz, Louis, 83.

Agricultural experiment stations, 86, notes 4 and 5.

Agricultural schools, growth of, as factor in applied research, 90.

Agriculture: Federal aid to research in, through land-grant colleges, 9, 22, 79; advances based on scientific research, 10, 11, 12; subject of study by the Bowman Committee, 73; distribution of undergraduate degrees among allied fields and, 177; Ph.D. degrees in, 177, 178.

Agriculture, Department of: Federal funds obligations for basic research by, xxv; interest in scientific research, 31-32, 83; establishment of, 84.

Air conditioning, 10.

Air Force, development of basic research programs by, xiii. See also Armed services; Military services.

Airborne infections, 53.

Allies: continued scientific training by, during World War II, 159; need for agreement with, on release of wartime-developed scientific data, 191.

Altitudes, development of devices to combat effects of, on airmen, 53.

Alumni associations, as inadequate source of medical research funds, 58.

American Association of Collegiate Registrars, distribution made by, of undergraduate degrees among academic and professional fields, and of Ph.D. degrees by science fields, 177-178.

American Council on Education, proposed utilization of the Committee on Measurement and Guidance of, in the scholarship and fellowship program, 181.

American Institute of Physics, calculations by, on deficits of scientific and technological personnel, 158.

American Universities, Association of: survey of universities and colleges accredited by, on expenditures for scientific research, 122 ff.

American Youth Commission, socio-economic study of students related to school attendance, 172, 173.

Anatomy: development of, basic to medical progress, xi, 14, 56; analysis of research in selected university departments of, 130; Ph.D. degrees in, 177; See also Research, medical.

195

Anderson, G. Lester, and Berning, T. J., *What Happens to High School Graduates*, data from, 171-172.

Anemia, 13, 54.

Antarctic Treaty, landmark in international scientific relations, xv.

Anthropology, in NSF support programs, xx; included within the scope of science considered by the Moe Committee, 142; distribution of Ph.D. degrees among other fields and, 177-178; Ph.D. degrees in, 178.

Applied Physics Laboratory, xiii.

Applied research. See Research, applied.

Archaeology, functional, in NSF support programs, xx.

Armed forces. See Armed services.

Armed Forces Institutes, need for scientific knowledge based on World War II to be made available in, 8; need for quality training in, 140, 161-162; work in, recommended as an allowance toward continued education; 165; proposed utilization of the examination staff of, in the scholarship and fellowship program, 181.

Armed services:
Dr. Bush on—
Need to salvage scientific talent in, during World War II, xvi, 7-8, 25-26; need for research to develop new weapons for, 9; need for professional partnership between civilian scientists and, 17-18; role of, in scientific research, 33-34; recommended representation of, in proposed National Research Foundation, 36.

Palmer Committee on—
Factors contributing to successes of medical departments of, World War II, 52; scientific interests of, as specialized, 62.

Moe Committee on—
Need to salvage scientific talent in, during World War II, 139-141, 158-162. See also Military services.

Stewart Committee on—
Need to salvage scientific talent in, during World War II, 189-190.

Army:
Dr. Bush on—
Role of, in medical research, 15; increased emphasis on science in officer training, 17; need for continued research by, 17-18; scientific students in, 24; as participant in wartime scientific research, 29; as board member to declassify scientific information, 29.

Palmer Committee on—
Death rate in, from disease, World Wars I and II, 49, 52; adoption of use of penicillin by, 53; role of, in medical research, 55, 56.

Bowman Committee on—
Cooperation with, as function of proposed National Research Foundation, 117.

Moe Committee on—
Recommended participation of, in program to locate and develop scientific talent in the armed forces, 140-141; students of medicine and engineering in educational programs of, 159; plans for integrated scientific training in, to reduce wartime deficit in trained personnel, 160-162, 165.

Stewart Committee on—
Wartime scientific research under auspices of, 190; need for participation of, in decisions on release of wartime-developed scientific data, 191.

See also Armed services; Army, Department of the; Military services; Talent.

Army, Department of the: development of basic research programs by, xiii. See also Armed services; Army; Military services.

Army Epidemiology Board, medical progress during World War II stimulated by, 52.

Army Medical Library, 118, 119.

"Army Universities" abroad, need for, 161-162.

Army's General Educational Development Tests, 165.

Arteriosclerosis, 14, 55.

Arthritis, 14, 55.

Associated Universities Incorporated, xxi.

Association of Universities for Research in Astronomy, xxi.

Asthma, 14,55.

Astronomy, Ph.D. degrees in, 177. See also Kitt Peak National Observatory; National Radio Astronomy Observatory; Associated Universities Incorporated; Association of Universities for Research in Astronomy.

Atabrine, 49, 53.

Atomic Energy, International Conference on the Peaceful Uses of, xv.

Atomic Energy Commission, program for medical research, xii; declassification of wartime scientific data by, xviii; Federal funds obligations for basic research for national defense by, xxv.

196

Attachés. See Scientific attachés.

Aviation, development of devices to combat effects of high altitudes, 53.

Aviation medicine, committee on, 53.

Bacteriology, development of, basic to medical progress, xi, 14, 56; analysis of research in selected university departments, 129, note 5 data, 130; Ph.D. degrees in microbiology and, 178. See also Research, medical.

Barton, Henry A., member, Committee on Discovery and Development of Scientific Talent, 44, 136.

Basic Research. See Research, Basic.

Battelle Memorial Institute, *Research in Action*, 86, note 3.

Baxter, J. P., III, member, Committee on Publication of Scientific Information, 45, 188.

Bell, Alexander Graham, 85.

Bernal, J. D., *The Social Function of Science*, 87, note 2.

Berning, T. J. See Anderson, G. Lester.

Bibliographic services. See Libraries.

Bids on research projects, recommendation to waive requirements on, 38-39.

Biochemistry: development of, basic to medical progress, xi, 14, 56; analysis of research in selected university departments of biology, including, 129; Ph.D. degrees in, 177.

Biological sciences, included within the scope of science considered by the Moe Committeee, 142; deficit in personnel trained in, 158; undergraduate degrees in allied fields and, 177; distribution of Ph.D. degrees among other fields and, 177-178; deficits in training research personnel in, 179.

Biology: development of improved course content for teaching, supported by NSF, xvii; term "medical research" to include related aspects of, 64, note 1; analysis of research in, 122-134, *passim*; Ph.D. degrees in, 178; graduate school enrollments in, 178.

Biology and Medicine, Division of, in NSF, xx.

Biophysics, analysis of research in selected university departments of biology, including, 129.

Blood, increased availability of plasma and, for transfusions, 49; fractionation studies, 49; plasma and substitutes, 53.

Book, William F., *The Intelligence of High School Seniors*, data from, 175-176.

Botany, Ph.D. degrees in, 178. See Biology.

Bowman, Isaiah, chairman, Committee on Science and the Public Welfare, 44, 71, 72, 73; transmittal of report, 71.

Bronk-Nesmeyanov Agreement, scientific cooperation under, xiv.

Brookhaven National Laboratory, xxi.

Bubonic plague, 53.

Buckley, Oliver E., member, Committee on Science and the Public Welfare, 44, 72.

Budget, Federal: amounts proposed for medical research, xi-xii; amounts proposed for military research, xii; for International Geophysical Year, xv; for NSF, xxiv-xxvi; proposals concerning, for proposed National Research Foundation; 39-40; estimated, for Federal aid to medical research, 60; suggested reforms in procedures, for scientific work of the Government, 100-101. See also United States Congress.

Budget, national research: public and private funds for research, 85-89.

Burdick, C. Lalor, member, Committee on Discovery and Development of Scientific Talent, 44, 136.

Bureau of Standards, work of, as example of Government's responsibility in background scientific research, 82.

Burns, studied, 53.

Bush, Vannevar, frontispiece; as Director of OSRD, vii; transmittal of *Science, the Endless Frontier*, vii, 1-2; letter from President Roosevelt, to, vii, 3-4; comparison of recommendations with developments since 1950, vii-xxvi.

Business administration, 177. See Social sciences.

California Institute of Technology, 140, 141; Jet Propulsion Laboratory of, xiii.

Cancer, ix, 13, 14, 54, 55-56.

Cardiovascular diseases, xi, 14, 55-56.

Carnegie Corporation, 84.

Carnegie Foundation: study by, in Pennsylvania, of able students in relation to college attendance, 170-171; proposed utilization of the Graduate Record Examination Office of, in the

197

scholarship and fellowship program, 181.

Carnegie Institution, 84, 86 note 6.

Castle, William B., member, Medical Advisory Committee, 43, 48.

Census, 1940, population under 20 years of age, 166; age and school attendance, 166-167.

Cerebral hemorrhage, 14, 55.

Chauncey, Henry, asst. sec., Committee on Discovery and Development of Scientific Talent, 44, 136.

Chemical companies, included in survey of research in industrial laboratories, 133, note 2.

Chemistry: development of, basic to medical progress, xi, 14, 56; development of improved course content for teaching, supported by NSF, xvii; future progress to be striking in, 78; analysis of research in, in universities, colleges, industrial research laboratories, and nonprofit science institutes, 122-134 *passim*; deficit in personnel trained in, 158, 179; distribution of undergraduate degrees in allied fields and, 177; Ph.D. degrees in, 177; graduate school enrollments in, 178.

Chemotherapy, committee on, 53.

Childhood, reduction in death rates in, various diseases, 54-55.

Cholera, 52, 53.

Civil Service, establishment of excepted category for scientific personnel, ix; reforms needed in, for scientific personnel, 7, 101-104; regulations of, to apply in proposed establishment of National Research Foundation, 35; separate branch of, recommended for scientific and technical positions, 76; up-grading of scientific positions recommended, 76.

Civilian controls in military scientific research, 33-34.

Civilian participation in declassification and publication of wartime scientific research data, 29-30, 191.

Coast and Geodetic Survey, establishment of, 84.

Coffey, Walter C., member, Committee on Science and the Public Welfare, 44-72.

Cold, common, 14, 55.

College attendance and graduation, statistics and analyses concerning, 166-176, *passim*. See also Tables.

College enrollment, statistics from studies in Indiana, Minnesota, and Pennsyl-

vania on mental ability, high school attendance and graduation, and on college enrollment, related to socio-economic groups, 144-145.

College Entrance Examination Board, proposed utilization of, in the scholarship and fellowship program, 181.

Colleges. See Universities and colleges.

Colleges, small, scientific research in, 123-124.

Commerce, Department of: Federal funds obligations for basic research by, xxv; interest in scientific research, 31-32.

Commerce, Secretary of: study by, concerning patent laws as they affect industrial research, 21.

Committee, Medical Advisory, membership, 43, 48; report by, 46-69.

Committee on Discovery and Development of Scientific Talent, membership, 44-45, 136; report by, 135-185.

Committee on Publication of Scientific Information, membership, 45, 188; report by, 186-192.

Committee on Science and the Public Welfare, membership, 43-44, 72; report by, 70-134.

Committees consulted by Dr. Bush, 43-45.

Communications company, included in survey of research in industrial laboratories, 133, note 2.

Compton, Arthur H., quoted, on deficit of trained scientific personnel, 159.

Compton, Karl T., member, Committee on Publication of Scientific Information, 45, 188.

Conant, James B., statement by, on development of scientific talent, 23; member, Committee on Discovery and Development of Scientific Talent, 45, 136; member, Committee on Publication of Scientific Information, 45, 188; as chairman of the National Defense Research Committee, 144; quoted, on future of science being dependent upon national educational policy, 144.

Congress. See United States Congress.

Contracts, research, need to waive certain statutory and regulatory fiscal requirements, 38-39. See General Accounting Office.

Convalescence, committee on, 53.

Cooperative Test Service, proposed utilization of, in the scholarship and fellowship program, 181.

Cox, Oscar S., member, Committee on

198

Science and the Public Welfare, 44, 72.

Davis, Horace Leonard, *The Utilization of Potential College Ability Found in June 1940, Graduates of Kentucky High Schools,* data from, 174-175.

Davis, Watson, member, Committee on Discovery and Development of Scientific Talent, 45, 136.

DDT, 13, 49, 52, 53.

Death rates, from disease, World Wars I and II, 5, 13, 49, 52; in childhood, 13, 14, 54-55; from disease, compared to war casualties, 1, 13-14, 54; reduction in, from various diseases, 54-55. See also Life expectancy.

Deaths, principal diseases causing, 14, 55.

Defense, Department of: peacetime vigor of, in research, xiii; declassification of wartime scientific data by, xviii; Federal funds obligations for basic research for national defense by, xxv.

Defense Mobilization, Office of, x.

Defense Science Board, xxii.

Deficiency diseases, 13.

Deficits in trained research personnel. See Talent.

Degenerative processes, 14.

Degrees, Ph.D.: deficits, 139; annually, in science and technology (6-year period), 150; proposed 300 fellowships for students working toward, 150-151; distribution of, by science fields, 177-178. See also Fellowships and scholarships.

Degrees, undergraduate: deficits, 139; in science and technology (1941), 150; proposed assistance to 6,000 students annually, to obtain, 150, 151; distribution of, by academic and professional fields, 177. See also Fellowships and scholarships.

Demography, in NSF support programs, xx.

Dentistry, term "medical research" to include related aspects of, 64, note 1; distribution of undergraduate degrees in allied fields and, 177. See also Biological sciences; Research, medical.

Dewey, Bradley, member, Committee on Science and the Public Welfare, 44, 72.

Diabetes, 13, 54.

Diet, 13, 54.

Diphtheria, reduction in death rate from, 54.

Disease, the "war" against, xi, 1, 3, 5, 7, 9, 10-11, 14-16, 49. See also Death rates; Diseases; Medical Advisory Committee; Research, medical.

Diseases, basic research as contributory to solutions to, viii-ix; emphasis shifted to those of middle- and old-age groups, 14, 55; principal death causing, 55; studies on, by Division of Medical Sciences, National Research Council, and Committee on Medical Research, OSRD, 53.

Diseases, childhood, reduction in death rate from, 54-55.

Diseases, deficiency, near eradication of, 54.

Diseases, infectious, committee on, 54.

Diseases, tropical, committee on, 53.

Divinity, distribution of undergraduate degrees in allied fields and, 177.

Doherty, R. E. member, Committee on Discovery and Development of Scientific Talent, 45, 136.

Doisy, Edward A., member, Medical Advisory Committee, 43, 48.

Drugs, committee on, 53.

Dykstra, Clarence A., member, Committee on Science and the Public Welfare, 44, 72.

Dysentery, 13, 49, 52, 53.

Earth sciences, distribution of Ph.D. degrees among other fields and, 177-178.

Ecology, human, in NSF support programs, xx.

Economic studies related to scientific research and its application, recommended as function of proposed National Research Foundation, 117.

Economics, in NSF support programs; xx; graduate school enrollments in, 178.

Economy. See National economy.

Edison, Thomas A., 85.

Education (as field of study), distribution of undergraduate degrees in allied fields and, 177. See also Social sciences.

Education, statistics on enrollments in schools at various levels, 26; able students lost to, at higher level, 166-176.

Education, scientific: five principles of Government support of, xx-xxi. See also Research, basic; Research, scientific; Scientific Personnel and Education, Division of.

Education, United States Office of: statistics on enrollments in public schools, 168.

Education and Labor, U. S. Senate Committee on: report by OSRD to a subcommittee of, on need for Federal support of medical research, 57.

Education and Training, NSF compilation of science policies on, xxiii.

Educational institutions, expenditures for scientific research, 85, 89. See also Research institutes; Universities.

Educational opportunity, intelligence related to, 173, Table III.

Educational policy, Dr. Conant quoted on future of science as dependent upon, xv, 23, 144. See also National science policy; Talent.

Educational programs, inadequacy of, for talented students, 169.

Educational pyramid (able students lost to higher education), studies concerning, 147-149, 166-176 (see also Tables); graduate students related to total college enrollments, 178.

Electrical companies, included in survey of research in industrial laboratories, 133, note 2.

Electronics, future progress to be striking in, 78.

Elicker, Paul E., member, Committee on Discovery and Development of Scientific Talent, 45, 136.

Employment, scientific progress related to, 6, 10, 11, 18, 74; promotion of research aimed at increasing basic scientific knowledge leading to new industries and increase in, as responsibility of proposed National Research Foundation, 116.

Encephalitis, 53.

Engineering, subject of study by the Bowman Committee, 73 (see also Committee on Science and the Public Welfare); costliness of research in, 79; deficit in personnel trained in, 158, 179; students of, in Army and Navy programs, 159; Ph.D. degrees in, 177; distribution of undergraduate degrees in allied fields and, 177.

Engineering, chemical: analysis of research in, 122-134, passim; Tables, 122, 128 note 4 data, 131, 133.

Engineering, electrical: analysis of research in, 122-134, passim; Tables, 122, 132, 133.

Engineering, sanitary: committee on, 53.

Engineering and natural sciences: Division of Mathematical, Physical and Engineering Sciences in NSF, xx; analysis of research in, and postwar needs by universities, colleges, industrial research laboratories, and nonprofit science institutes, 122-134, passim; Tables, 122, 127-133.

Engineering Education, Measurement and Guidance Project in: proposed utilization of, in the scholarship and fellowship program, 181.

Engineering schools: growth of, as factor in applied science, 90; recommended grants to, for industrial research services, 108.

Entomology, term "medical research" to include related aspects of, 64, note 1; economic, 129 note 5 data; Ph.D. degrees in, 178.

Europe, former source of basic scientific knowledge, 6, 22, 78; use of public funds in, for medical research, 50, 56-57; governmental support of science in, prior to support in U. S., 83.

Executive Orders: 10512, xii; 10521, xxiii; 10807, xxiii.

Experiment stations, agricultural research in, 32. See also Agriculture.

Federal aid for research. See Research, basic; Research, industrial; Research, medical; Research, scientific.

Federal Council on Science and Technology, xxii.

Federal Financial Support of Research Facilities, NSF compilation of science policies on, xxiii.

Federal Funds for Science, annual report of NSF, xxiii-xxiv.

Federal Security Agency, interest in scientific research, 31-32.

Fellows, proposed means and procedures for choosing, 180-185. See also Scholars; Fellowships and scholarships.

Fellowships, proposed for medical research, xi, xii; as recommended form of Federal aid to medical research, 50-51, 59-60, 66-68; program of, supported by Rockefeller Foundation, administered by the Medical Fellowship Board of the National Research Council, 60; postdoctoral research, recommended as means of increasing trained scientific personnel, 97-98; senior research, recommended for mature investigators as means of increasing scientific personnel, 98; international, recommended as part of needs for international scientific cooperation, 114; scholarships, grants-in-aid and, as normal means of developing leadership, 143; proposed plan for, as means of assistance to doctoral students in science, 150-157, *passim*; responsibilities of educational institutions to provide training commensurate intellectually with superior ability, 152; State quotas for, 155; machinery for administration, 156-157; tenure of, and conditions of continuance to a Scholar or Fellow, 156-157; eligibility of Scholars for, 157. See also Fellowships and scholarships; Talent.

Fellowships, National Research Council, 156.

Fellowships and scholarships:
Highlights of proposal for a program of, to renew scientific talent—
Summary of proposals and the program as it has developed under NSF, xv-xvi; budgets for fellowship program, xvi; nonimplementation of scholarship program, xvi.

Dr. Bush on—
Need for a Government agency to administer, 9; summary of Moe Committee recommendations for, as means of scientific training, 26-27; to be administered by proposed National Research Foundation, 34, 35,' 37, 38, 39-40.

Bowman Committee on—
Recommended, to be administered by proposed National Research Foundation, 75, 117.

Moe Committee on—
Proposed Federal program for, 137, 138-139; financial need not to be a factor in granting of, 152; recipients of, to constitute a National Science Reserve, 153; obligations of recipients of, 153-154; method of selection, 154-155; estimated cost of selection process, 185.

Fibrin foam, 53.

Fine arts, graduate school enrollments in, 178.

Fiscal procedures, U. S. Government: suggested reforms in, for scientific work of the Government, 100-101. See also General Accounting Office.

Foreign Policy, Science and, Department of State, xiv.

Forestry, distribution of undergraduate degrees in allied fields and, 177. See also Biological sciences.

Forrestal, James V., Secretary of the Navy, interest of, in postwar military research, 159.

Foundations, endowments from, as partial sources of funds for medical research, 58.

Frank, Lawrence K., secretary, Committee on Discovery and Development of Scientific Talent, 44, 136.

Franklin, Benjamin, influence upon American science, 83.

Furer, Admiral J. A., quoted on need for military research, 159.

Gas gangrene and casualties, 53.

General Accounting Office, need to relax certain requirements of, for research contractors, xxii, 38, 39.

Genetics, 129, note 5 data; Ph.D. degrees in, 178.

Geography: economic and social, in NSF support programs, xx; included within the scope of science considered by the Moe Committee, 142; graduate school enrollments in, 178.

Geological Survey, establishment of, 84.

Geology, included within the scope of science considered by the Moe Committee, 142; deficit in personnel trained in, 158, 179; Ph.D. degrees in, 177; graduate school enrollments in, 178.

Germany, 143.

G. I. Bill of Rights, summary of impact of, xvi, xvii; recommendation that scale of support for scientific scholarships and fellowships be same as that provided by, 138, 152; place of, and recommended changes in, for ameliorating scientific and technological deficits, 140, 141, 162-163. See also Veterans Administration; Veterans Readjustment Assistance Act.

Glass company, included in survey of research in industrial laboratories, 133, note 2.

Goetsch, Helen B., socio-economic study by, of able students related to college attendance, 172-174.

Goodpasture, Ernest, member, Medical Advisory Committee, 43, 48.

Government, U. S., responsibilities in fields of scientific research and development:

Highlights of Dr. Bush's recommendations and developments since 1950—Science, the Endless Frontier as classic expression of desirable nature of relationships between science and the Government, vii; role of the Government and developments in that role since 1950 in promotion of basic scientific research, viii-x; increase in Federal funds for research, ix; role of the Government in industrial research, x-xi; role of the Government in medical research, xi-xii; role of the Government in military research, xii-xiii; role of the Government in promoting international scientific cooperation, xiii-xv; role of the Government in promoting the discovery and developemnt of talent, xv-xvii; role of the Government in publication of scientific knowledge developed during World War II, xviii-xix; the National Science Foundation as the means of the Government's playing its role in scientific development, xix-xxvi; five principles of support by the Government for scientific research and education, and highlights of NSF compliance with, xx-xxi; appropriations and obligations for basic research, xxv-xxvi.

Dr. Bush on—
Aid to scientific research by public and private organizations, 1, 3; need for support to medical research, 5-6, 14-16; need for support of scientific research, 6-7; competition with industry and universities for scientific personnel, 7; need for support in developing scientific talent, 7; need for incentives to industry to conduct research, 7; summary of responsibility in scientific development, 8-9; recommendation to establish in the permanent Government structure an agency to carry out Government responsibility, 9; advancement of science as the concern of the Government, 11-12; fundamentals underlying support for scientific research, 12; responsibility for scientific military research, 17-18; increase in applied scientific research, 1930-1940, in the Government, 19-21; recommendation for improving personnel policies concerning scientific personnel employees, 20; scientific research in the Government as essentially applied research, 20; recommendation that basic scientific research at colleges, universities, and research institutes be strengthened by use of public funds, 20; recommendation for creation of a Science Advisory Board

to coordinate policies and budgets of Government agencies engaged in scientific research, 20-21; means of strengthening industrial research, 21; role of the Government in promoting international flow of scientific information, 22; role of the Government in supporting basic scientific research, 22; support to research in Agriculture, 22; cost of adequate support to basic and applied scientific research, 22; mobilization of science for World War II, 28-29; various agencies of the Government as participants in wartime scientific research, 29; responsibility to make available the results of wartime research data, 29-30; responsibilities of the Government in promotion of scientific research and scientific talent, 31-40; mechanism for and fundamentals of the Government's role in scientific research and development, 31-40; need for a special agency to assist scientific research outside the Government and to support research on weapons and administer a program on science scholarships and fellowships, 31-32; five fundamentals in a program for Government support for scientific research and education, 32-33.

Palmer Committe on—
See Research, medical: Palmer Committee.

Bowman Committee on—
Role of the Government in conducting scientific research, 73; role of the Government in promotion of research in fields of natural sciences, engineering, and agriculture, 73; role of the Government to encourage international exchange of scientific knowledge and engineering art, 73, 75; role of the Government to carry on scientific research within the Government, 75-76; recommendations for increased effectiveness of scientific work in the Government, 76; achievements made by Federal research agencies, 77; responsibility of the Government to promote scientific progress in the national welfare, 77; precedents in Federal aid to scientific progress, 77, 78, 79; need for Federal aid to scientific research in private institutions, 77-80; responsibility of the Government to support pure scientific research, 81-82; responsibility of the Government to support background scientific research, 82-83; creation of Federal scientific bureaus, 83-85; Government support of the Wilkes Exploring Expedition, 84; expenditures for scientific research, 85-89; necessity for support to pure research in universities in order to maintain proportion of pure to applied

research, 87, 88-89; recommended form of support to research in universities, 93-98; nature of scientific research in the Government and suggested reforms, 99-106; recommended aids to industrial research and technology, 107-109; desirable role of the Government in international scientific cooperation, 113-114.

Moe Committee on—
See Talent, scientific, programs for discovery and development of: *Moe Committee.*

Stewart Committee on—
Responsibility of the Government to speed release of scientific information developed during the war, 187-192.

Government-Industry Relationships on Research, NSF compilation of science policies on, xxiii.

Government - University Relationships in the Conduct of Federally Sponsored Research, NSF compilation of science policies on, xxiii.

Grants and grants-in-aid:
Highlights of proposals for, and developments since 1950, xi, xii, xx-xxi.

Palmer Committee on—
As means of advancing medical research, 50-51, 60, 63, 66-68.

Bowman Committee on—
As means of assistance to research, 95, 96, 97, 108, 116-117.

Moe Committee on—
Scholarships, fellowships, and, as normal means of developing leadership, 143.

Griffiths, Farnham P., member, Committee on Discovery and Development of Scientific Talent, 45, 136.

Harvard University Library, 118.

Haskins, C. P., member, Committee on Science and the Public Welfare, 44, 72.

Havighurst, Robert J. See Warner, W. Lloyd.

Hay fever, 14, 55.

Health, scientific progress essential to, 5, 74, 77; promotion of research for improvement of, as responsibility of proposed National Research Foundation, 116. See Research, medical; Wartime Health and Education, Subcommittee on, of U. S. Senate Committee on Education and Labor.

Hemolytic streptococcal disease, 53.

Henry, Joseph, 83.

High schools: inadequacy of teaching of science in, and actions of NSF since 1950 to improve teaching in, xv, xvii; studies concerning able students lost to higher education, 144-145, 166-176; need for means for veterans and wartime industry workers to complete education in, 163-165, *passim.*

Historical studies related to scientific research and its application, recommended as function of proposed National Research Foundation, 117.

History, graduate school enrollments in, 178.

Home economics, distribution of undergraduate degrees in allied fields and, 177.

Horticulture, Ph.D. degrees in, 178.

Hospitals, Federal funds obligations for basic research in, xxv; contracts with, by Committee on Medical Research, as wartime measures, 53; place of, in medical research, 56. See also Research, medical.

Hunter, W. S., member, Committee on Discovery and Development of Scientific Talent, 45, 136.

Hygiene, advances in, 13; in World War II, 49, 52.

Immune globulins, 49, 53.

Indiana, study on socio-economic groups of high school students related to ability, and on intelligence levels of students planning to go to college, 144-145.

Indirect Costs, NSF compilation of science policies on, xxiii.

Industrial medicine, 53.

Industry and industrial progress:
Highlights of problems and Dr. Bush's recommendations—
Basic research as fundamental for, viii, 19; need for more basic research by industry, ix, 22; NSF surveys of industry's research and development efforts, xxiii; dependence upon trained scientists, 7, 9; need to strengthen patent system as incentive to, 7; need for scientific knowledge based on problems of World War II to be made available, under controls, to industry, 8; role of industry in applied science compared with basic research, 22; incentives to be maintained for research by industry, 31.

Palmer Committee on—
Industry as partial source of funds for medical research, 58.

Medical supplies, committee on, 53.

Medicine:
    Dr. Bush on—
    OSRD program on problems of, in wartime, xii; advances in, World War II, made possible by backlog of scientific data, 13.

    Palmer Committee on—
    Contribution to World War II by, 49, 52-53; committees on aviation and industrial medicine, 53; achievements during World War II under direction of OSRD, 53-54; studies during World War II by Division of Medical Sciences of the National Research Council and Committee on Medical Research, 53-54; civilian medicine as roots of wartime medicine, 54; emphasis shifted to middle- and old-age groups, to malignant diseases and degenerative processes, 55; importance of fundamental research to progress, 55-56; scope of sciences related to medicine, 56.

    Moe Committee on—
    Students of medicine in Army and Navy programs, 159; distribution of undergraduate degrees in medicine and allied fields, 177; Ph.D. degrees in surgery and medicine, 178.

    See also Medical Advisory Committee; Research, medical.

Medicine, Division of Biology and, in NSF, xx.

Medicine, Welch Fellowships in, 98.

Meningitis, 13, 49, 52.

Mental illness, 5, 14, 55.

Metallurgy, Ph.D. degrees in, 177.

Meteorology, Ph.D. degrees in, 177; graduate school enrollments in, 178.

Microbiology, Ph.D. degrees in bacteriology and, 178.

Military research. See Research, military.

Military services: programs for medical research, xii; continuance of research programs after World War II and expansion from applied to basic research, xiii; programs for continuing education of highly qualified men, xvi. See also Armed services; G. I. Bill of Rights; National defense; Research, military.

Milwaukee, Wisconsin, study of able high school graduates by socio-economic status related to college attendance, 172, 173, 174, 175.

Minerology, Ph.D. degrees in, 177.

Mines, Bureau of, establishment of, 85.

Mining, 177.

Minnesota, studies on high school students and graduates entering college, related to socio-economic groups and mental ability, 145, 171-172.

Moe, Henry Allen, chairman, Committee on Discovery and Development of Scientific Talent, 44, 136; transmittal of report, 136.

Moe Committee: Dr. Bush's excerpts from report by, on development of scientific talent, 23, 24-25; Dr. Bush's summary of program proposed by, 26-27; scope of science as considered by, 142. See Committee on Discovery and Development of Scientific Talent.

Moulton, Harold G., member, Committee on Science and the Public Welfare, 44, 72.

Municipal governments, need for Federal cooperative support of research by, 101.

Music, graduate school enrollments in, 178.

Naples epidemic, typhus, 52.

National Academy of Sciences:
    Activities since 1950 pertinent to Dr. Bush's recommendations—
    Role in promoting international scientific conferences, xiv; science attaché program urged by, xiv; role of, in the International Geophysical Year, xiv-xv.

    Dr. Bush on—
    Establishment of Research Board for National Security by, 33, 34; recommendations as to relationship of, to proposed National Research Foundation, 35-36.

    Bowman Committee on—
    Report to Congress by, on Government scientific activities, 1884 and 1908, 99; cooperation with, as function of proposed National Research Foundation, 117.

    Moe Committee on—
    Proposed role of, in a scholarship and fellowship program, 139, 154, 156, 180-181; scope of science within the purview of, 142.

    Stewart Committee on—
    Recommended as agency to control release of wartime-developed scientific data, 191-192.

    See also National Research Council.

National Advisory Committee for Aeronautics: Federal funds obligations for

207

Bowman Committee on—
Recommended creation of, policies, functions, and basic structure of, as Federal agency to promote science in the interest of public welfare, 75, 115-117; recommendations concerning relation to universities, 94-98; recommendation that a proposed Science Advisory Board cooperate with, 106; suggested responsibility of studying technological development in industry and of experimenting with methods of aid to industrial research, 107; recommended power to make grants to universities, engineering schools, and nonprofit industrial research institutes, 108; proposed investigation by, of advisability of Federal aid to encourage new scientific enterprises of applied-research nature, 109; recommended functions of, in international scientific cooperation, 113-114; proposed functions related to improvement of library services, 118-121.

See also National Science Foundation; National Scientific Research Foundation.

National resources:
Dr. Bush on—
Scientific progress related to conservation of, 10-11.

Bowman Committee on—
Conservation of, as dependent upon scientific progress, 74, 77; promotion of research aimed at conservation and better utilization of, as responsibility of proposed National Research Foundation, 116.

Moe Committee on—
Intelligence of citizenry included among, 137, 144.

National Resources Committee, Research —A National Resource, 86, notes 2, 4; 100 and note 1.

National Roster, inventory by, of graduate students in nonprofessional and non-vocational schools and departments, by science and other fields, 178.

National Science Board: convictions of, concerning an undergraduate program of scholarships, xvi; approval of establishing an Office of Social Sciences in NSF, xx; high calibre of staff, xxi.

National Science Foundation: Act of 1950 creating, vii, xiii; resumé of accomplishments related to Dr. Bush's recommendations for a National Research Foundation, vii-xxvi; establishment of Division of Biological and Medical Sciences, xi, xii; promotion of international conferences by, xiv; science attaché program urged by,

xiv; funds obligations of, compared to Dr. Bush's proposals on funds requirements, xvi-xvii; support of research directed at new approaches to information problems and translation techniques, xviii; scientific information activity of, xviii-xix; establishment of a Science Information Service and a Science Information Council, xviii, xix; highlights of early and present organization, present functions and relationships, and comparison with Dr. Bush's proposals, xix-xxvi; relationships of the Director with other Federal agencies and committees, xxii; annual report, Federal Funds for Science, xxiii-xxiv; Federal appropriations for, xxiv-xxvi. See also National Research Foundation; National Scientific Research Foundation.

National science policy: need for, and the National Research Foundation as focal point for development of, in Dr. Bush's recommendation, vii, 12, 31, 34; active growth of, since 1950, vii-xxvi, passim; role of NSF in, xxii-xxiv; NSF compilations on, xxiii.

National Science Reserve:
Dr. Bush on—
Proposals for, 27.

Moe Committee on—
Proposed establishment of, and recommendation that recipients of science scholarships and fellowships be enrolled in, 138, 153-154; to include veterans trained under the G. I. Bill, 163; suggested administrative organization, bases of selection, and procedures for selection of scholars and fellows, 180-185.

National Scientific Research Foundation: proposed establishment of, and functions in the program for discovery and development of talent, 180-185, passim. See also National Research Foundation.

National security:
Dr. Bush on—
Scientific progress essential to, 5; need for civilian-controlled organization, in liaison with Army and Navy, with appropriated funds to initiate military research, 6; scientific research related to, 17-18.

Bowman Committee on—
Scientific progress essential to, 74, 77.

Stewart Committee on—
Release of wartime-acquired scientific information related to, 190.

See also National defense; Research Board for National Security; Research, military.

National welfare:
Science related to, since 1950—
Increased awareness of relationships between science and, vii-xxvi, *passim.*

Dr. Bush on—
Science related to, vii-viii, 5, 6-7, 9, 10-12, 17-22, 40; scientific talent related to, 23; need for coordination of research programs in interest of, 31.

Bowman Committee on—
Scientific progress essential to, 74, 77.

Moe Committee on—
Science as a member of a team in, 142-143; training of highest ability students essential to, 146; promotion of public interest in development of science as important to, 157; discovery and development of scientific talent in the armed forces, and special postwar educational programs for veterans and wartime industrial workers as essential to, 158-165; adverse effects of deficits in trained scientific personnel on, 158-165, *passim;* recruitment by industry of students who should continue education as adverse for, 169-170.

Stewart Committee on—
Release of scientific information developed during the war as important to, 187-192.

Natural resources. See National resources.

Natural sciences:
Dr. Bush on—
Proposal for a division of, in a National Research Foundation, 35, 39; support of research in, as function of proposed National Research Foundation, 35, 37, 39-40.

Bowman Committee on—
Subject of study by the Committee, 73; costliness of research in, 79; expenditures for research in, 86, note 5; analysis of research in engineering and, and postwar needs by universities, colleges, industrial research laboratories, and nonprofit science institutes, 122-134, *passim.* See also Committee on Science and the Public Welfare.

Moe Committee on—
Undergraduate degrees (1941) in technology and, 150; Ph.D. degrees (6-year period) in technology and, 150; statistics on research in technology and, 177.

Naval Observatory, establishment of, 84.

Naval Research Laboratory, establishment of, 85.

Navy:
Dr. Bush on—
Role of, in medical research, 15; increased emphasis on science in officer training, 17; need for continued research by, 17-18; scientific students in, 24.

Palmer Committee on—
Adoption of use of penicillin by, 53; role of, in medical research, 55-56.

Bowman Committee on—
Cooperation with, as function of proposed National Research Foundation, 117.

Moe Committee on—
Recommended participation of, in program to locate and develop scientific talent in the armed forces, 140-141; students of medicine and engineering in educational programs of, 159; plans for integrated scientific training in, to reduce the wartime deficit in trained personnel, 160-162, 165.

Stewart Committee on—
Wartime scientific research under auspices of, 190; to participate in decisions on release of wartime-developed scientific data, 191.

See also Armed services; Army; Navy, Secretary of the; Navy Department; Navy Postgraduate School.

Navy, Secretary of the: joint statement with Secretary of War, to National Academy of Sciences, regarding scientific progress and the national security, 17; release of results of wartime medical research to the public, 28.

Navy Department: support to basic research, xiii; as participant in wartime scientific research, xiii, 29; as board member to declassify scientific information, 29. See also Armed services; Military services.

Navy Postgraduate School, xvi.

Neuropsychiatry, 53.

Neurosurgery, 53.

New England Industrial Research Foundation, as a research clinic, 107-108.

New York, State of: Board of Regents of, quoted on imperativeness of higher education for ablest students, 144; "Regents' Plan" for postwar education of demobilized armed forces and wartime industry workers, 164-165.

New York Public Library, 118.

Norcross, Cleveland, secretary, Committee on Publication of Scientific Information, 45, 188.

Nuclear energy, international aspects of diversion of, into constructive uses, xv.

Nuffield College, University of Oxford, 151.

Nursing, 177.

Ochsner, Alton, member, Medical Advisory Committee, 43, 48.

O'Donnell, J. Hugh, member, Committee on Science and the Public Welfare, 44, 72.

Office of Scientific Research and Development: Dr. Bush as director of, vii; Dr. Bush's recommendation that portions of wartime research programs should be carried on in peacetime, xii, 18; continuation by the military services of research arrangements initiated by, xii-xiii; functions and emergency nature of, 3, 33; wartime scientific research by, 12, 190; scientific mobilization by, 28-29; publication of results of wartime research by, 29-30; research sponsored by, on nonprofit basis, 39; broad authorities to, and relaxation of technical fiscal procedures concerning research contracts, 39; medical advances attributable to Committee on Medical Research of, 49, 52; cost of medical program to July 1944, 49, 54; organization of Committee on Medical Research under, as wartime measure, 53; scope of activity in medical research, 53-54, 55; report to U. S. Senate on need for Federal support of medical research, 57; costs of contracts with colleges and universities, 1943-44, 87; decisions in early days of, on release of medical-research data, 189; publication plans of, concerning release of wartime-developed scientific data, 192.

Oil companies, included in survey of research in industrial laboratories, 133, note 2.

Organization and Administration of Research, NSF compilation of science policies on, xviii.

Osborn, General Frederick H., letter from Dr. Moe to, quoted on need for scientific training for armed forces, 161.

Oxford, University of (Nuffield College), 151.

Paleontology, Ph.D. degrees in, 177.

Palmer, Walter W., chairman, Medical Advisory Committee, 43, 47, 48; transmittal of committee report, 47.

Parasitology, development of, basic to medical progress, xi, 14, 56.

Paratyphoid, 52.

Parsons, Dr. Charles L., quoted on waste of scientists in World War II, 159.

Patents:
Dr. Bush on—
Need for strengthening patent systems, to assist small industries, 7; problems concerning patent laws bearing upon industrial research, 21; policy on, related to operations of proposed National Research Foundation, 38.

Palmer Committee on—
Question of rights on discoveries made under Government - sponsored research, 61.

Bowman Committee on—
Impact of system on research by industry, 76; needed strengthening of the system, 109; recommendation concerning policies of proposed National Research Foundation, 117; reference to policy concerning universities, research institutions, and recipients of grants, 117; policy concerning inventions by Government employees, 105.

Pathology: development of, basic to medical progress, xi, 14, 56; committee on, 53.

Pauling, Linus, member, Medical Advisory Committee, 43, 48.

Pearl Harbor, 139.

Penicillin, 10, 13, 14, 49, 52, 53, 55.

Pennsylvania, studies on high school graduates and college attendance of high-ability students, 145, 170-171, 172, 173.

Pepper, Senator Claude D., quoted, on Government's role in sponsorship of research, 63.

Perazich, G., and Field, P., *Industrial Research and Changing Technology*, 86, note 2.

Personnel, scientific:
Highlights of recommendations concerning scientific personnel in the Government and developments since 1950, ix.

Dr. Bush on—
Recommendations concerning, in the Government, 7.

Bowman Committee on—
Recommendations concerning, in the Government, 76, 101-104.

Moe Committee on—
Deficit in, due to wartime interruptions to education, 150; deficit in technological and scientific personnel

from war and selective service policies, 158-160.

See also Civil Service; Talent.

Pharmaceutical industry: role of, in the war on disease and medical research, 13, 15; included in a survey of research in industrial laboratories, 133, note 2.

Pharmacology: development of, basic to medical progress, xi, 14, 56; Ph.D. degrees in, 178. See also Table V, 130, note 4.

Pharmacy, 177.

Physical science: Division of Mathematical, Physical and Engineering Sciences in NSF, xx; support of research in, as function of proposed National Research Foundation, 35, 39-40; institutions for research in, included in a survey of research in nonprofit science institutes, 133, note 3; included within the scope of science considered by the Moe Committee, 142; undergraduate degrees in mathematics and, 177; distribution of Ph.D. degrees among other fields and, 177-178; graduate school enrollments in, 178; deficits in training research personnel in, 179.

Physical Sciences Study Group, Massachusetts Institute of Technology: NSF's support of, to improve science teaching, xvii.

Physicians, role of, in the war on disease, 13.

Physics: development of, basic to medical progress, xi, 14, 56; development of improved course content for teaching, supported by NSF, xvii; analysis of research in, in universities, colleges, industrial research laboratories, and nonprofit science institutes, 122-134, passim; deficit in personnel trained in, 158, 179; Ph.D. degrees in, 177; graduate school enrollments in, 178; studies by Institute of, on deficits in training of research personnel in physical sciences and engineering, 179.

Physiology: development of, basic to medical progress, xi, 14, 56; analysis of research in selected university departments, 129, 130; Ph.D. degrees in, 178. See also Research, medical.

Plant pathology, 129, note 5 data.

Plastics, 10.

Pneumonia, 13, 49, 52, 53, 54-55.

Population: increase, 1900-1940, 10-11; statistics on child and youth, related to school attendance, 166-176. See also Tables.

Primary schools, loss of talent in, 147.

Private organizations: President Roosevelt's question on the role of the Government to aid scientific research activities by, 1, 3, 73, 77; need for Federal cooperative support of research by, 101. See Committee on Science and the Public Welfare.

Prizes, establishment of, as function of proposed National Research Foundation, 117.

Protozoology, 64, note 1.

Psycholinguists, included in NSF support programs, xx.

Psychology: social, included in NSF support programs, xx; included within the scope of science considered by the Moe Committee, 142; deficit in personnel trained in, 158, 179; distribution of Ph.D. degrees among other fields and, 177-178; Ph.D. degrees in, and graduate school enrollments in, 178.

Public, need to promote interest of, in scientific development, 157.

Public health: role of groups engaged in, in the war on disease, 13; distribution of Ph.D. degrees among other fields and, 177-178.

Public Health Service, U. S.: source of funds for medical research, xii; role of, in medical research, 15, 55, 56; scientific interests of, specialized, 62. See also Health.

Public organizations: President Roosevelt's question on the role of the Government to aid scientific research activities by, 1, 3, 73, 77; need for Federal cooperative support of research by, 101. See also Committee on Science and the Public Welfare.

Public welfare. See Committee on Science and the Public Welfare; National welfare.

Publication. See Committee on Publication of Scientific Information.

Publications and Scientific Collaboration, Division of, proposed in a National Research Foundation, xxiv, 35, 39. See also National Science Foundation; Science Information Service, Office of.

Rabi, I. I., member, Committee on Science and the Public Welfare, 44, 72.

Radar, 10, 17.

Radiation Laboratory, M. I. T., xiii.

211

Radiation treatment for cancer, 13. See also Cancer.

Radio, 10.

Rayon, 10.

Reference aids, functions of NSF in providing, xviii, xix. See also Libraries.

Refractory diseases, xi, 14.

"Regents' Plan for Postwar Education in the State of New York," 164-165.

Rehabilitation, committee on, 53.

Renal disease, xi, 14.

Research—A National Resource, National Resources Committee, 100 and note 1.

Research: functions of NSF in publishing results of, xviii.

Research, applied:
Highlights of the Report and developments since 1950—
Lack of desirable balance between Federal funds for basic research and, ix; research programs of military services expanded beyond, to basic research, xiii; distinguished from pure research, xxvi; as driving out pure research, xxvi.

Bowman Committee on—
Nature of, 83; tendency to drive out pure research, 83; relative expenditures, U. S. and England, for pure research and, 87; need to maintain balance with pure research, 88-89; advent of, in universities, 90.

See also Research, industrial; Research institutes.

Research, basic:
Highlights of the Report and developments since 1950—
Development of national policy concerning, since 1950, vii-viii; as scientific capital, viii; as pacemaker of technological progress, viii; as fundamental to industrial progress, viii; National Research Foundation as focal point for support of, in Dr. Bush's recommendations, viii; as principal focus of the Report, viii-ix; increase in use of Federal funds for, ix; lack of desirable balance between Federal funds for, and applied research and development, ix; necessary to national defense, ix; necessary to research training, ix; necessity for, by industry, ix; need for industry to support, in colleges and universities, ix; research programs of military services expanded to include, xiii; average amounts and duration of NSF grants for, xx; permission for grants for, to vest title to research equipment, xxi;

NSF compilation of science policies on, xxiii; widening support of, by various Federal agencies, and Federal appropriations and funds obligations for, xxv-xxvi; recognition of importance of, revealed by Federal legislative actions and appropriations, xxv-xxvi; ratio of Federal basic-research funds to over-all research and development funds, xxvi; distinguished from applied research, xxvi; tendency to be driven out by applied research, xxvi.

Dr. Bush on—
Need for a Government agency to supplement, in colleges, universities, and research institutes, 9; colleges, universities, and research institutes as centers of, 12, 19, 20; backlog of scientific data accumulated through, as basis of advances in medicine, World War II, 13; in the war against disease, 13-15; need for public funds to strengthen, 20; support of, by Government, as an aid to industrial research, 21; role of Government in promoting, 22, 31; fundamentals in use of public funds for, 32-33; support of, as function of proposed National Research Foundation, 34, 38, 39-40.

Bowman Committee on—
Need to improve transition between, and industrial application, 75, 78; as pacemaker of technological progress, 78-79; nature of (pure and background scientific research), 81-83; Government's responsibility to support, 81-83; tendency for applied research to drive out, 83; relative expenditures, U. S. and England, for applied research and, 87; need to maintain balance with applied research, 88-89.

Moe Committee on—
Recruitment of talented students by industry for applied science as a deterrent to expansion of talent in, 149; need to maintain appropriate balance between those trained in scientific fields and those trained in social sciences, arts, and humanities, 179; deficit in training personnel in physical sciences and engineering, 179.

See also Research, scientific.

Research, biological: institutions for, included in a survey of research in nonprofit science institutes, 133, note 3.

Research, industrial:
Highlights of Dr. Bush's analysis of factors related to, x-xi.

212

required in a program for Federal aid to, 79, 80; essentials of a program for Federal aid, 80; recommended establishment of a National Research Foundation to administer Federal aid, 80; Federal support of, in universities and nonprofit institutes recommended, 80; components and nature of, 81-83; development of, in U. S., 83-85; national (public and private) expenditures for, 85-89; basic elements in a national policy for, 88-89; in American universities and colleges, 90-98; operating costs for, related to salaries of research staffs in universities and colleges and in industrial research laboratories, 92-93; need to assist colleges and universities with operating costs, 93; recommended forms of Government support in universities, 93-98; nature of, in the Government and suggested reforms, 99-106; bearing of Federal income tax upon expenditures for development and, 110-111; recommended legislative action on income-tax laws designed to aid, 111-112; responsibility of proposed National Research Foundation to promote, 116; analysis of expenditures for, by 125 universities and colleges, and postwar needs, 122-124; example of opposition to Federal aid for, 124; detailed analysis of expenditures for, in a small sample of leading universities, industrial research laboratories, and nonprofit science institutes, 125-134.

Stewart Committee on—
Effect of the war on, and need to release wartime-developed data as soon as possible, 189-192.

See also Research, basic.

Research Board for National Security: establishment of, 33; recommended participation of, in program to locate and develop scientific talent in the armed services, 140-141, 160.

Research clinics, advantages to small business, 107-108.

Research facilities, scientific and technical: NSF compilation of science policies on Federal support of, xxiii; responsibility of proposed National Research Foundation to provide for, where inadequate, 116.

Research in Action, Battelle Memorial Institute, 86, note 3.

Research institutes: NSF surveys of research and development efforts of, xxiii; Federal funds obligations for basic research in, xxv; as centers of basic research, 6, 12, 19, 20; expenditures for scientific research, 6-7, 85-89; need for public funds to

strengthen, 20; inadequacy of expenditures for basic research by, 22; fundamentals of public-funds support of research in, 33; support of scientific research in, as function of proposed National Research Foundation, 34, 37, 38, 39-40; role of the Government to assist research in, 73, 74, 75, 101; achievements made by, 77; recommended grants to, for industrial research and for fundamental research, 108; reference to patent policies of, 117.

Research workers, assistance to, as function of proposed National Research Foundation, 117.

Richards, A. N., member, Committee on Publication of Scientific Information, 45, 188; chairman, Committee on Medical Research, OSRD, 57; quoted, on deficiency in funds of universities for medical research, 57.

Rockefeller Foundation, 60, 84.

Rockefeller Institute of Medical Research, 84, 86, note 6.

Rodent control, 53.

Rogers, Walter S., member, Committee on Discovery and Development of Scientific Talent, 45, 136.

Roosevelt, President Franklin D.: letter to Vannevar Bush from, requesting recommendations on means of scientific advancements, 3-4; Dr. Bush's reply, 1-2; references to the President's letter, 13-14, 54, 73, 77, 137, 142, 187, 189, 190, 192; letter to, from Dr. Charles L. Parsons, quoted, on waste of scientists in World War II, 159.

Russia, recommendation for appointment of scientific attachés at embassy in, xiii-xiv.

Russian sputnik, effects on activity of the Government in scientific field, x, xxv.

Ryerson Laboratory of the University of Chicago, 140, 161.

Scarlet fever, reduction in death rates from, 54.

Scholars, proposed means and procedures for choosing, 180-185. See also Fellows; Fellowships and scholarships.

Scholarships: nonimplementation of proposed program for, and substitutes therefor, xvi; precautions necessary not to syphon a disproportionate amount of ability into science from other fields through, 142-143; fellowships, grants-in-aid and, as normal means of developing leadership, 143;

proposed plan for, as means of assistance to undergraduates in science, 150-157, *passim*; responsibilities of educational institutions to provide training commensurate intellectually with superior ability, 152; State quotas for, 155. See also Fellowships and scholarships; National Defense Education Act; National Merit Scholarship Corporation.

School attendance, statistics on, 166-176. See also Tables.

Schools, graduate: NSF support for renovation and equipment of research laboratories of, xxii; need for policies of, to assist veterans in making up lost time in scientific training, 163-165. See also High schools; Primary schools; Secondary schools.

Science:
Highlights of the Report and developments since 1590—
Summary of Dr. Bush's recommendations on the role of the Government in the development of science and progress since 1950, vii-xxvi; lack of full awareness of difference between technology and, ix; international relations in, since 1950, xiii-xv; widening of international relations in, to include political considerations, xv; inadequacy of teaching in, as found by Dr. Bush, and actions of NSF since 1950 to improve teaching, xv-xvii; NSF projects on history and philosophy of, xx; NSF annual report, *Federal Funds for Science*, xxiv; technology distinguished from, xxvi.

Dr. Bush on—
Scope of, as referred to by President Roosevelt, 1; in the war against disease, 1, 3; mobilization of, for World War II, 28-29.

Palmer Committee on—
Scope of, related to medicine, 56.

Bowman Committee on—
Status and trends in, in America, 81-89.

Moe Committee on—
Scope of, within purview of National Academy of Sciences, and of the report of the Committee, 142; inadequacy of teaching of, in secondary schools, 148-149; statistics concerning training of personnel for technology and, undergraduate and graduate schools, 177-179.

See also Defense Science Board; Federal Council on Science and Technology; Research, applied; Research, basic; Research, medical; Research, scientific; Science, applied.

Science, applied:
Dr. Bush on—
Function of proposed National Research Foundation to improve transition from research to, 37, 39-40.

Bowman Committee on—
Planned coordination and direction as benefits to, 79-80; methods of progress in, during war, not feasible in peace, 80.

Moe Committee on—
Recruitment of talented students by industry for, as a deterrent to expansion of talent for basic research, 149.

Stewart Committee on—
Emphasis on, during war, 189.

See also Research, applied; Research, scientific.

Science, Department of: recommended creation of, by the National Academy of Sciences, 99.

Science, Federal Funds for, NSF annual report, xxiv.

Science Adviser to the Secretary of State, establishment of position, xiv.

Science Advisory Board: recommendations for establishment of, to coordinate scientific work of Government agencies, x, 7, 20-21, 76, 105-106.

Science Advisory Board (temporary), appointed by President Roosevelt, 100.

Science Advisory Committee, the President's: operations of, related to Dr. Bush's recommendations, x, xiv, xviii; science attaché program urged by, xiv; NSF relationships with, xxii.

Science and Astronautics, House Committee on, vii.

Science and Foreign Policy, Department of State, xiv.

Science and Technology, Department of: debate as to need for establishment of, related to Dr. Bush's recommendations, x.

Science and Technology, Federal Council on: recent establishment of, related to Dr. Bush's recommendations, x.

Science and Technology, Special Assistant to the President for: establishment of position of, x; NSF relationships with, xxii.

Science Information Council, establishment of, by NSF, xviii.

Science Information Service, Office of, NSF, xviii, xix, xx, xxiv, note 4.

215

Science policy. See National science policy.

Science Reserve. See National Science Reserve.

Scientific attachés: recommendation for appointment of, in certain embassies, viii-xiv, 114; development of the program for, since 1950, xiii-xiv.

Scientific capital: basic scientific research as, viii, 6; drawn upon during the war, 189.

Scientific collaboration. See Publications and Scientific Collaboration, Division of.

Scientific cooperation. See International scientific cooperation.

Scientific enterprises, unresolved question by the Bowman Committee as to Federal aid to launching of, 108-109.

Scientific information: release of, developed during the war, xviii, 186-192; activity of NSF in dissemination of, xviii-xix, 75, 116; NSF compilation of policies on, xxiii.

Scientific Personnel and Education, Division of, NSF, xx; proposals for, 35, 39.

Scientific progress, significance of, in all phases of American life, 10-12.

Scientific publications, encouragement of, as function of proposed National Research Foundation, 35, 37, 39-40.

Scientific research. See Research, scientific.

Scientific talent. See Talent.

Scientific training: need for, related to public welfare, 6-7; English scholars quoted on need for and nature of, at undergraduate level, 151-152. See all Research entries; Talent.

Scientists, need for professional partnership between the military services and, 17-18; assistance to, as function of proposed National Research Foundation, 117. See all Research entries; Talent.

Secondary schools: loss of talent in, and variation of quality of education in, 147-149; need to assist able students to complete education in, 149. See also High schools; Schools; Talent.

Security restrictions on wartime scientific data, coordination required in lifting, 29-30. See also Scientific information.

Seismology, Ph.D. degrees in, 177.

Selective Service policies, effect of, on deficit in scientists, and recommended corrective action, 139-141, 158-160, 163.

Serum albumin, 49, 53.

Shapley, Harlow, member, Committee on Discovery and Development of Scientific Talent, 45, 136.

Shock, studied, 53.

Silliman, Benjamin, 83.

Slichter, Sumner, x.

Small business, research clinics as advantage to, 107-108.

Small Business Act of 1958, x.

Smallpox, 52.

Smith, Homer W., secretary, Medical Advisory Committee, 43, 48.

Smithson, James, bequest for Smithsonian Institution, 84.

Smithsonian Institution: Federal funds obligations for basic research, xxv; creation of, 84.

Social Function of Science, The, Bernal, 87, note 2.

Social sciences: need to maintain balance between talents in natural sciences, medicine, and, xx, 23; undergraduate degrees in, 177; statistics on research in, 177; graduate school enrollments in, 178; warning not to drain too many able students from research in, for scientific research, 179.

Social Sciences, Office of, NSF, xx, xxi.

Socio-economic studies: relationships between socio-economic status, ability, and level of education attained, 171-176; relations between parental income and college courses pursued, 174.

Sociology, in NSF support programs, xx.

Soviet Professional Manpower, National Research Council, xxv.

Space. See Aeronautical and Space Sciences, Senate Committee on; National Advisory Committee for Aeronautics; National Aeronautics and Space Council; Science and Astronautics, House Committee on.

Sproul, Dr. Robert Gordon, President, University of California: quoted, on the citizenry's intelligence as the most valuable national resource, 144.

Standard of living: scientific progress related to, 10-12, 18, 74, 77; need for program to locate and develop scientific talent in armed services as basic to postwar conditions of, 141.

217

for Government aid, 7; recommendations for development in armed services, 7-8; deficit in talent, 7-8, 18, 23-24; need for Government scholarships and fellowships, 8; development of talent by Government as an aid to industrial research, 21; nature of problem, 23-27; Government responsibility, 31; development of talent as function of proposed National Research Foundation, 34, 38, 39-40.

Bowman Committee on—
Need for talent for research in the Government, 76; as responsibility of proposed National Research Foundation, 116-117.

Moe Committee on—
Federal responsibility, 137-185, *passim;* long-term plans (early schooling of more able students, scholarships and fellowships), 137-139, 147-157; plans for the near future (to locate and develop talent in armed forces), 137, 139-141, 158-165; precautions necessary, not to syphon a disproportionate amount of ability from other fields, 138, 142-143, 145, 150; scope of science used by the Committee, 142; suggested means and procedures for discovery and development, 180-185.

Stewart Committee on—
Development of, in armed forces, necessary before demobilization, 189-190.

Talented students: statistics from studies in Indiana, Minnesota, and Pennsylvania on mental ability, high school attendance and graduation, and college enrollment, related to socioeconomic groups, 144-145, 147-148; studies concerning able students lost to higher education, 166-176.

Tate, J. T., vice chairman, Committee on Science and the Public Welfare, 44, 72.

Tax laws: effect of, on industrial research, x, 76, 110-111; recommended legislative action on, designed to aid research and development, 111-112. See also Internal Revenue Code.

Taylor, Hugh S., member, Committee on Discovery and Development of Scientific Talent, 45, 136.

Technical schools, need for policies of, to assist veterans in making up lost time in scientific training, 163-165.

Technological progress, basic research as the pacemaker of, viii, 19, 78-79.

Technology: lack of full awareness of difference between science and, ix; science distinguished from, xxvi; recommended Federal aids to, 107-109; undergraduate degrees (1941) in natural sciences and, 150; Ph.D. degrees (6-year period) in natural sciences and, 150; statistics on research in natural sciences and, 177. See also Federal Council on Science and Technology.

Tests: in the program for discovery and development of talent, 180-185, *passim.* See American Council on Education; Carnegie Foundation; College Entrance Examination Board; Cooperative Test Service; Engineering Education, Measurement and Guidance Project in; Armed Forces Institute; Iowa, University of.

Tetanus, 13, 49, 52.

Thomas, Dr. Charles Allen, quoted on need for scientific training to replace technical men in the armed forces, 159.

Thomson, Elihu, 85.

Transfusions, blood, committee on shock and, 53.

Translation of scientific data, NSF functions in providing, xviii, 116. See also Libraries.

Truman, President Harry S., xix.

Tuberculosis, 53, 54-55.

Turner, Kenneth B., assistant secretary, Medical Advisory Committee, 43, 48.

Tuve, M. A., member, Committee on Publication of Scientific Information, 45, 188.

Typhoid, 52.

Typhus, 13, 49, 52, 53.

U-boats, battle against, as battle of scientific techniques, 6, 17.

Ulcers, peptic, 14, 55.

United States: place of, in discovery of fundamental new knowledge and basic scientific principles, 78; progress in applied science, 78; development of scientific research in, and growth of Federal participation in research, 83-85. See Government, U. S.

United States Congress: Increased attention to scientific research and development and legislation and appropriations conducive thereto, vii-xxvi, *passim;* standing committees in, concerned with science and technology, viii; need for appropriations by, for scientific research in Government to be assured on long-term basis, 76. See also Budget, Federal.

Veterinary medicine, 64, note 1.

War, resources of proposed National Research Foundation to be available in event of, 117; recommendation for establishment of a National Science Reserve to be available upon declaration of, 138, 139, 153. See also National Science Reserve.

War, Secretary of: joint statement, with Secretary of the Navy, to the National Academy of Sciences, regarding scientific progress related to the national security, 17; release of results of wartime medical research to the public, 28; Research Board for National Security established at request of, 33.

War Service School, recommended, for special training of veterans and former workers in war industries, 165.

Waring, James J., member, Medical Advisory Committee, 43, 48.

Warner, W. Lloyd, et al., Who Shall be Educated: The Challenge of Unequal Opportunities, data from, 172-174.

Wartime Health and Education, Subcommittee on, of U. S. Senate Committee on Education and Labor: reports of OSRD to, on need for Federal support of medical research, 57.

Washington, George, plan for a national university, 84.

Weapons, need for a Government agency to support research on, 9. See Research, military.

Weaver, Warren, member, Committee on Science and the Public Welfare, 44, 72.

Weed, Lewis H., chairman, Division of Medical Sciences, National Research Council: quoted, on need for Federal support for medical research, 58.

Welch Fellowships in Medicine, 98.

What Happens to High School Graduates, G. Lester Anderson and T. J. Berning, data from, 171-172.

Who Shall be Educated: The Challenge of Unequal Opportunities, W. Lloyd Warner, et al., data from, 172-174.

Whooping cough, 54.

Wilkes Exploring Expedition, Federal support of, 84.

Wilson, Carroll L., member, Committee on Publication of Scientific Information, 45, 188.

Wilson, E. B., member, Committee on Discovery and Development of Scientific Talent, 45, 136.

Wilson, Robert E., member, Committee on Science and the Public Welfare, 44, 72.

Wistar Institute, 86, note 6.

Wood, Ben D. See Learned, W. S.

World War II, progress of medicine during, xi, 49, 52-53; contributions to scientific knowledge during, 1, 3; reduction in death rate from disease from rate in World War I, 5, 13; recommendation that scientific knowledge developed during, be made available, 8, 28-30; advances in medicine during, made possible by backlog of scientists and scientific data, 13, 49, 52; period between World War I and, marked by U.S. leadership in medical research, 15; scientific developments during, 17; deficit in scientific students as result of, 24, 25, 139, 150, 158-160; mobilization of science during, 28; effect of, on additions to basic medical research, 49, 54; effect of, on scientific training and research, 91-93; 189-192.

Wrather, William E., member, Committee on Science and the Public Welfare, 44, 72.

X-ray, 53.

Yellow fever, 13, 49, 52.

Zoology, Ph.D. degrees in, 178.

Made in the USA
San Bernardino, CA
19 November 2014